Fundamental Aspects of
Appropriate Technology

Fundamental Aspects of Appropriate Technology

Proceedings of the International Workshop on
Appropriate Technology

Center for Appropriate Technology
Delft University of Technology
September 4 - 7, 1979

J. de Schutter/G. Bemer/editors

Delft University Press
Sijthoff & Noordhoff International Publishers
1980

Published by

Delft University Press
Mijnbouwplein 11
2628 RT Delft
The Netherlands

and

Sijthoff & Noordhoff
International Publishers
P.O. Box 4
2400 MA Alphen aan den Rijn
The Netherlands

Cover by Jan van Wessum, Amsterdam

ISBN-13: 978-90-286-0640-1   e-ISBN-13: 978-94-009-9139-2
DOI: 10.1007/ 978-94-009-9139-2

Contents

Session 3: The Framework for Appropriate Technology.
The Role of Appropriate Technology in (Rural) Development

Preface

Between 4-7 September 1979, an international workshop on Appropriate Technology (AT) was organized in Delft, Netherlands, by the Center for Appropriate Technology of the Delft University of Technology. Representatives of 24 AT organisations from all over the world held discussions on the role of AT as a factor in development.

There were two main objectives of the workshop :

- to enlarge the understanding of, and knowledge about the processes and conditions essential for the introduction of AT in regional development programs. This was formally referred to as 'the implementation of the results of AT research'.

-secondly, an evaluation of the theories and models which have been applied for the establishment of these regional development programs. This was formally referred to as 'an inventory of AT concepts.'

The workshop discussions focussed essentially on three issue areas: technology and development, organisational framework, and education and research. A summary of the conclusions and recommendations made by the workshop can be found in Chapter One of this report of the proceedings.

All participants were invited, prior to the workshop, to outline their ideas on the subjects listed above, in position papers. Condensed versions of these papers are presented in Chapter Three.

We would like to point out that, although all participants gave permission for publication, the contents of this report of the proceedings remain the full responsibility of the editors. This applies in

particular to the conclusions and recommandations as well as to the
statements included in the edited transcripts of the discussion
highlights of the presentation of the papers during the Workshop.

Our thanks are due to the participants of the workshop, who collectively contributed to the intensive discussions; to the two Ministerial
departments of the Netherlands Government responsible for Science
Policy, and Development Cooperation, and to the Delft University of
Technology for their financial sponsorship of the workshop; and to the
workshop staff, for their responsible and total contribution to the
workshop organisation.

The editors

I. Conclusions and Recommendations

Technology and development

Of these two concepts development is the more important one. Development is
defined here as a process of socio-economic change of a society. This process should be directed towards various development goals, e.g., the satisfaction of basic human needs, social participation and control and ecological soundness. From the assumption that development itself is the central
variable in any activity in a society, it follows that technology can only
be seen as a means to achieve development goals. Before starting any technological activity there must be principle agreement on the development
goals to be achieved. Hence, the characteristics of appropriate technology
are directly related to and deducted from development theory and practice.
It is in this context that one must decide wether or not "small is beautiful" or wether appropriate technology will be "technology with a human
face", "technology of limitation", "soft technology" or other widely used
descriptions.
It is insufficient to state that any technology meeting the needs of a
society is appropriate technology. Guidelines for appropriate technology
activities have to be developed, whereby it is of major importance that the
activities are geared towards the basic needs of the neediest of the world.
In particular this concerns the people who live in the rural areas of developing countries and in the slum areas of cities in both developing - and
industrialized countries. In short the poor are the main target for appropriate technology, which obviously gives appropriate technology a political
dimension too.

Appropriate technology is however a matter of both theory and practice. The
importance of "doing things" was very positively demonstrated by the contributions presented in session 1 of this Workshop. Contradictory to certain

1

critiques it was demonstrated that succesful applications of appropriate technology have been realized in various parts of the world. It was emphasized furthermore that any activity, however marginal in terms of national development aims, can mean the difference between life and death for the people involved. This will always be an important underlying factor in the setting up of appropriate technology projects.

When considering the technology aspect of appropriate technology one has to bare in mind that, during the last decades, countries and regions within countries have not been technologically isolated. Western countries and developing countries became linked more and more and it is therefore that appropriate technology has to deal with two aspects of major importance : the technology introduced by activities of industrialized countries and the originally existing technological tradition. It is expected that a combination of these two aspects may give appropriate technology an important impact on development.

A key element in the setting up of appropriate technology projects is the comprehensive knowledge of all aspects involved. These include culture, sociology, economy, ecology, technology and management. Appropriate technology is therefore a multidisciplinary activity in which technology itself might only contribute 10% to 30% of the solution of the problem. There can be no doubt however as to the quality standards of a technological solution. The technology applied should always be properly designed and reliable.

Appropriate technology and economy

The possible impact of appropriate technology on economy has often been discussed. In the past its potential for job-creation was highlighted. In the papers and discussions of the third session the limitations to the importance of appropriate technology as a factor in economy were summarized as follows :

1. the immense magnitude of the development problem.

2. the technical potential for labour-intensive techniques has been largely exaggerated.

3. institutional constraints for development and implementation of appropriate technology on the regional level have been largely ignored.

4. the choice for a development model is basically a political decision
   made on the national level.

Hence, when discussing the role of appropriate technology in development we
have to distinguish between decision making processes on the national and
regional levels.

At present most of the economic theory and practice have been developed for
an open economic system. This system implies the production for international markets. As a consequence there is an increasing need for international
division of labour. In the discussions of the third session also two other
economic systems were distinguished.

- the inward directed economic system.
This means production mainly for the country itself with very little emphasis on the exportation of goods.
- autarky, which is a strategy of disassociation.

It was felt that appropriate technology would be most valuable to an inward
looking economic system. Further it was pointed out that one should be very
careful in the application of established economic theories to future development strategies in third world countries. This does not mean that many
elementary economic rules can be ignored in appropriate technology activities. In this context the issue of commercial viability has been discussed.
There can be no arguing about the fact that any appropriate technology project should be properly managed and should have a sound financial basis.

National and regional level policy

On a national level, main directives for appropriate technology projects may
be their impact on such issues as : self reliance, income distribution and
sectoral priorities. The realisation of development aims on the national
level is hence more a matter of social organisation than involvement in the
development of technology hardware. Given the limitations mentioned previously, it was felt that future development of developing countries can only
partially follow the path of industrialisation as has been the case in many
western countries. The majority of the labour force will have to be absorbed
by the agricultural sector. In the manufacturing sector highly capital -
intensive methods are often used. In this way often intersectoral productivity differentials are created. Governments should see to it that these,
do not become reflected in wide income differentials.

3

On the regional level, technology policy should be mainly directed towards the definition of local needs and the technology to fulfil these needs. Local governments and local entrepreneurs will be mainly concerned with the development and production of appropriate technology hardware and implementation.

Motivation and participation

Perhaps the most important element that appropriate technology can contribute to development is the direct participation of all people involved in a process of technology choice and implementation. Motivation and participation hence are keyfactors in appropriate technology activities. A truly democratic system for decision making will therefore enhance appropriate technology activities. This introduces a strong socio-cultural element in appropriate technology. The process of participation should start at the lowest decision making level in a community : the family. Moreover it is important to notice that, when talking about rural development, approximately 51% of the people involved are women. Until now too little emphasis has been given to their role in appropriate technology.

Appropriate technology characteristics

It is still not possible to give a uniformly applicable definition of appropriate technology, but this might not be important. The most important result of the appropriate technology debate until now, may be that it at least gives an opportunity to eliminate technologies which are certainly inappropriate against the background of community development.
Summarizing appropriate technology may be characterized as follows :

1. normatively, in the sense of a value, appropriate technology means that we do things that enable people to benefit from it directly.

2. strategically it means the use of a model for technological development which enables people to rely on their own capacities.

Local organisations

When discussing the organisational framework for appropriate technology it should be realized that activities in this field are all directed to the fulfilment of basic needs of basically unpowerful people. Their development goals are related to a development strategy in which self-reliance is the main factor. Local organisations must therefore always be the key-actors. in project activities. If participation and motivation of these people are not assured it might be better not to start any activity at all. Every appropriate technology should therefore start with the initiative of such a local organisation. This Workshop therefore recommends to support local appropriate technology organisations and their activities.

Infrastructure

Appropriate technology projects require the construction of an appropriate organizational framework. It appears that non-conventional organisations, who are very open to new ideas, are the best instruments for the execution of the task. They should start to organise people to work in small entreprises and in well defined programmes. Also the setting up of regional field stations and development centers can be important contributions to the introduction of appropriate technology. The work could be very well coordinated on a regional level. In this way there would be a cross fertilization of knowledge and skills of people working on the regional level and the abilities of the people directly involved in technology development on the project level. A large obstacle may be the attitude and training of the people on the regional and national level. Many of the decision makers and scientists are not familiar with the problems in the rural areas. Their attitude is often a rather technocratic one. An effort should be made to change that attitude, e.g. by changes in the contents and the approach of their training.

The setting up of an organizational framework for - and the execution of often very complicated development programmes can only be done by a long term approach. People involved in appropriate technology projects should be given the possibility to commit themselves for a period of at least five to ten years. The work should be done in a very professional way with the involvement of as many disciplines as necessary. Also a strong relationship between educational institutes and appropriate technology groups should be established. In this way field experience could directly and continouesly influence the educational programs in schools, universities and research institutes.

Evaluating education and research programmes against the background of demands formulated in appropriate technology, some essential shortcomings can be detected. These relate to the present understanding that technology is a factor in development. Educational programmes should be analysed following the statement that "technology is too important to let only engineers deal with it".

Students should be trained to work in multidisciplinary teams. In this way they are confronted with important factors in development work, e.g. sociology, economy, politics, ecology, culture and the various methods for introduction and implementation of technological solutions. This is a vast but challenging task for educational institutions in industrialized countries as well as in the developing countries.

Research programmes involving appropriate technology should first of all have a long-term character.

Important issues such as technology policy, economics of technology, traditional technology and need assesment methodology studies should be included. Principally these programmes should be of such a character that they can respond to - or even directly influence national policy changes. Furthermore it is important that there is a good working relationship between universities, research centers and appropriate technology groups operating in the field. Important aids for the establishment of such a relationship can be short term missions for training and advise, a network for exchange of documentation and litterature and provisions for easy and fast communication.

The role of research centers will be limited when it comes to design and construction. Development of appropriate technology hardware will be mainly a task for local organisations. Many institutions however are currently doing relevant work in the development of appropriate technology gadgets. Often this work can be seen as supplementary to and supportive of field research programmes.

Session 1: Topics in Appropriate Technology Projects

THE GAVIOTAS PROGRAM**

*Jorge Zapp Glauser**

I come from what we call "a Project" in Latin America. This project really is the center of action in a region. It is the region of the Savanna and the whole area drains into the Orinoco river. It has a size of approximately 300.000 km^2 and a population of only 20.000 people. Into the region there is a pressure from the mountains and from the populated areas of Colombia.

As an initiative from within the region, the geografical center started to create a model for development which would be sound in terms of ecological criteria in the first place. This is very critical in our case because we have a very stable ecological system, but at the same time a very fragile one. May be it is even one of the most fragile ecological systems in the world. At the same time a second objective was to settle as much people as possible in that region. The models which we developed may now be used for about two million people. At the moment we are working with five to six thousand families. However the center I am speaking about is mainly directed to the development of software i.e. : education, health, agricultural systems, etc., I am going to speak mainly about some of the hardware results we developed and compatible with the restrictions of this region. What we really want to do here is to develop a settlement model which is constantly one step ahead of the settlement itself. We do this with only a small group of people because every study shows, and we are also experimenting this very fast, that there will be a strong migration into this free but very fragile land.

*I.D.C. - Las Gaviotas, Bogota, Colombia

**Gaviotas is spanish for Seagull. In english literature the Gaviotas-program is also known as "The Case of the Seagulls".

THE PROJECT

We are working in a few big planes in the Northern part of Latin America.
There are old rivers which all drain into the Orinoco river. There are severe
erosion problems which force people into the cities and into our region. The
region as a whole is part of the migration studies of the Orinoco planes.
There are savannas which are many times tortured by fire; fire is a very
important exological factor. So it is possible that in the middle of the
savanna all grass and all living species are burnt during a dry season.
Furthermore there are small creeks which carry out water to the regions covered with jungle. These sustain some 90 to 95 percent of the living species
here and are the key to the ecological stability of the whole system.
Gaviotas is settled approximately 500 km from the nearest paved road and is
at a distance of about 500 km from the Venezuela frontier. In the middle
of the region we have a concentrated service station. People served by
Gaviotas live at a distance of sometimes 300 to 400 kilometers from this
center. Gaviotas has a central hospital, a central school, central supply
shops, central meteorological services, etc.. We think that education is
a key factor in development and have therefore developed our own educational
system. In many cases education is done in the field; not only for children,
but also for adults.

Gaviotas is involved in many and various types of hardware development activities* Among these are :

- production of vegetables in artificial soil

- breeding of cattle and african sheep

- workshops and local architecture

- production of soil-cement blocks (CINVA-ram)

- growing of palmtrees and palm-oil extraction

- wind measurement program and building of windmills especially for low
  wind velocities

- building of dams

* A U.N.D.P.-film, giving a very clear impression on various projects now
  going on at Gaviotas was made in 1978. The film may be obtained via
  U.N.D.P., NEW YORK, USA.

- building and installation of water turbines (e.g. Mitchell - and asher - turbine)

- construction of pumps using the water flow as a driving force

- modification and production of waterrams

- extraction of starch from maniok (cassava)

- building of a hand mill for sugar cane processing

- growing of Carribean pine trees and African grass

- installation of bio-gaz plants

- use of a plastic film tube fore concrete pipeline construction

- use of solar energy for water heating purpose

ORGANISATION

Gaviotas is a strange type of organisation. It looks governmental on Monday, Wednesday and Friday. The rest of the week it is a private, non profit, foundation. It was created by the initial settlers of the region. They asked the government to give them a supply center, which should have a central hospital, a health center, a small school, some stores and communication facilities. Only later on these people started to find out that just supplying things was not enough. Therefore, about 10 years ago, people from this center started to look for people from universities, government, industry, etc. and tried to absorp them into the center. This way these people became part of the region.

Today, Gaviotas is almost a self supporting project, because it sells part of it's technology to the cities in order to finance the development of the region. An example of this is the installation of solar water heating systems in the city of Bogota done by Gaviotas technicians. We receive help from government, international organisations, etc., but always on our terms. Gaviotas sometimes even hires experts for some specific type of problem and pays for them. May be the most important aspect of the project is that people are very proud of what is obtained and of what they are obtaining at the moment. One key element in this is that all innovations come from the users. This continuous evaluation has lead to a technology that every time looks more and more like a very sophisticated and reliable solution than it looks like old tyres and pieces of wood. In the beginning we thought very much in terms of lowest costs and simple to understand. Now the people have forced us to think

11

in terms of the most reliable solution. Always however we can say that this solution costs at least from four to ten times less than any alternative developed untill now. There is also always a message behind these alternatives. The message not only one of self reliance, but also of protection of an environment which is not only in our case but also in general a very prevalent factor.

STATEMENTS

- The main element in development is the family. In this family the women
  play the most important role and this fact should never be underestimated.

- Perhaps the most important factor in development is that solutions can not
  just be figured out. These have to evaluate from the people themselves.

- It is impossible to transfer appropriate technology gadgets, which have
  been produced in one place, to another place where these should be implemented. The reasons for adaptation vary very much from one place to another in a very complicated process.

- It is important to study traditional technology and let no knowledge about
  this disappear.

AN ANALYSIS OF FACTORS AFFECTING THE SUCCESSFUL CONTINUITY OF A SOLAR
DISTILLATION PLANT IN THE WEST INDIES

*Tom A. Lawand**

BACKGROUND

Source Philippe is a small fishing and farming community on the island of
La Gonave, about 50 kilometres off the coast of mainland Haiti. Prior to an
influx of mainland during the 1940's and early 1950's, much of the island
was covered with the lush vegegation and dense undergrowth typical of many
tropical areas. However, immigration quickly depleted the land of its one
vital resource, wood. Wood was used to produce charcoal, which was exported
in vast quantities to fuel the cooking fires of the mainland population.

By the late 1940's the tree-cover of the island was noticeably reduced, ex-
posing much of the mountainous and sloping surfaces to rain and wind erosion.
By the early 1950's vast stretches of bed rock has been exposed and significant climatic changes occurred. Rainfall, which between 1931 and 1946 had
averaged 1250 mm per year, had fallen to below 100 mm by 1957.

The reduced rainfall and erosion of the land had obvious detrimental effects
on the island's agrarian population as farming became increasingly incapable
of providing them with a livelihood. The more than 250 residents of Source
Philippe practiced less-than-subsistence farming during 3 or 4 months a year.
The result was a poverty stricken economy which at best could be described as
marginal.

Fishing practises were crude, but nevertheless they still reduced the coastal
fish population to the extent that the catch was insufficient to support even
a fraction of the community.

*B.R.I., Faculty of Engineering Mc Donald College of Mc Gill University
Quebec, Canada*

Some cattle raising on the arid land provided minimal amounts of protein food for a time. However, most of the animals were sold on the mainland to buy other cheaper and much needed food as well as clothing. The desperate situation in the village was somewhat alleviated by several aid programs which provided food-for-work projects such as road building. Otherwise, starvation would have been widespread.

There are no permanent rivers or streams on the island; the only source of potable water is provided by a few wells and springs. In the village of Source Philippe there is a well, but the brackish water was only used to water livestock. Each family had to rely on sending someone on a daily nine hour round-trip,by foot or donkey, to the nearest fresh water source. During extended periods of drought, these sources often dried up and the villagers were forced either to drink the contaminated water from their own well, or to depend on passing fishing boats to bring fresh water from the mainland. Drinking their brackish wellwater brought sickness to the villagers but it was impossible to rely on water from the fishing boats, because often weeks would pass before a boat would stop at the village.

The problems facing the residents of Source Philippe can be briefly summarized as follows :

1) Poor agricultural production has resulted from inadequate irregular rain-fall, deforestation, soil erosion, and poor agricultural techniques.

2) Health problems have been caused by the lack of potable water, lack of medical services and malnutrition.

3) Major communication and transportation difficulties have not only created problems for marketing agricultural produce, but isolation has also re-sulted in a lack of interest in the island by government and other potential sources of economic assistance.

4) Lack of education, either classical or practical, has made it difficult for the people to cope with their changing environment.

5) Economic means are barely adequate for survival.

6) A "culture of poverty" has developed following years of exploitation of the majority by the few more powerful members in the community, and in response to a physical environment which made it even more impossible for the poor to obtain any benefit from their labour. Daily survival was the

major preoccupation of the villagers.

In 1965 a rehabilitation project was begun in the region of Source Philippe.
The project was sponsored by l'Eglise Methodiste d'Haiti, a mainland church,
under the leadership of a Haitian agronomist who was familiar with the village
and with the island of La Gonave as a whole. The emphasis of the project was
on improving agricultural techniques to help the islanders cope with the new
climate and soil conditions. An agricultural cooperative was formed with the
aim of providing better farming techniques through practical education, group
credit, and the maintenance of adequate reserves to cover crop failures.
Large rainwater catchment and storage tanks were built to provide fresh water
for the village. A medical dispensary and artisans' workshop were also built
and roads to adjacent villages were begun. A fishing cooperative was started
to enable fishermen to purchase equipment capable of extending their fishing
range to the deeper off-shore waters.

These activities have had various degrees of success. Hunger and malnutrition
have become less crucial problems. Farming practises have improved, without
the introduction of any new tools, and the improvement in the quantity and
quality of agricultural produce has enabled the village and surrounding area
to become nearly self-sufficient. The effects of the climate have been less-
ened somewhat by new food and water storage practices, but long period with
no rain could bring the village to the brink of disaster.

PROBLEM

An extended period of drought during 1967 and 1968 showed that the rainwater
catchment and storage system was not an adequate long term solution to the
fresh water problem. The reforestation program which had been started would
not significantly improve ground water retention for one or possibly two generations. Drilling for fresh water in the surrounding hills had been tried
but without success. One possible solution proposed by the local leadership
was to find a way to extract fresh water from sea water, or from the brackish
and polluted well water.

SOLUTION

The choices available for water desalinization included mechanical and thermal separation processes which involve the use of relatively complicated machinery. It was recognized that if such equipment were installed the local population would become dependant on outside help to operate and maintain it and that they would have to buy increasingly more fuel to run the machinery.

It was obvious that the initial capital investment for the system would have to be found outside the community because the people were not in a position to make even a modest financial contribution. It was also recognized that any recurring costs for the operation of the system should be kept to an absolute minimum so that when the project terminated the community could assume sole responsibility for operating costs. Local conditions therefore suggested that an alternative to complex mechanical or thermal desalinization processes should be employed.

A solar distillation system was selected because it met the specific needs of the community. Most of the construction materials and all the skills for construction, operation and maintenance were locally available and there would be no need to import fuel to operate the plant. In addition, a solar still is non-polluting, as it operates quietly and emits no odors. The by-product of its operation is a residue of slightly more concentrated sea or brackish water, which can either be returned to the original source or allowed to further evaporate until all that remains is the salt.

Financial support was found for the project and the Brace Research Institute of McGill University undertook the preliminary design work, in consultation with the local leadership. A project engineer was assigned to provide technical assistance during the construction phase. Site selection and preparation were started in January 1969, and by June of the same year the plant was in full operation. The local population participated, either directly or through its leadership, in all aspects of the project from its inception through site selection, design, modifications and construction.

Construction of the installation was labour intensive and most members of the community, including men, women and children, were involved at one time or another. Much of the work, such as land levelling and carrying of sand, stones water and concrete, had to be done manually, so there was ample work for everyone.

17

Various levels of skills and capabilities were necessary throughout the project. Carpenters were required to build concrete forms. Bricklayers and masons were needed for construction of the solar still basins. Basic plumbing and tinsmith work was also required. All of these skills were available to varying degrees within the village. In addition, several community members became involved in educational programs related to the concept of solar distillation and to the operating and maintenance procedures for the plant.

Locally available materials were used as much as possible. The only items imported into the village were the glass, the rubber basin liner, and the sealing compound to hold the glass in place.

Briefly, a windmill is used to pump brackish water from a well up to a slightly elevated water storage tank. The brackish water then flows by gravity into the solar still basins, where the water is heated by the sun's radiation. As the brackish water is heated, pure water evaporates and condenses on the underside of the glass, which is cooler than the air inside the still. The condensed water flows down the under face of the sloping glass into a gutter. The pure water then flows by gravity through the gutter and into a fresh water storage tank.

The solar still operates best when the sun is shining but it does function at lower output under overcast conditions. The solar still in Source Philippe has also been adapted to serve as a rain catchment.

RESULTS

Soon after it was completed, the solar still in Source Philippe was producing an average lf 1250 litres of fresh water per day from brackish and sea water. This output is more than fifty percent greater than orginally planned. The system has been designed to permit its expansion as water needs increase. This can be accomplished easily and inexpensively by constructing additional evaporating basins.

The community does not require outside skills or imported fuel to operate the still. It has achieved an independence in being assured of a potable water supply which is not affected by periods of drought. The community has also developed generally by participating in the project; it has broadened its range of experiences in the process of using its own skills and familiar working materials.

In order to evaluate the project, a sociological study was undertaken eighteen months after the solar still was completed. The study concluded that the plant did not adversely affect the traditional social patterns of the village. The greater volume of collected and stored water which is available for daily use by the villagers means that they now spend less time filling this need and more labour is available to work in the fields. In addition, the people have insurance against the periodic droughts which the region experiences.

It is difficult to separate the effects of the solar distillation plant from the effects of other projects which are going on in the community. For example, health conditions of the people have dramatically improved and this has quite definitely been linked to the building of a medical dispensary, improvement in the quality and quantity of food available as well as to the steady supply of clean potable water from the solar still.

The solar still on the island of La Gonave has now been in operation for eleven years. There have been a few technical problems but all essential maintenance has been handled by members of the community. Adequate provision of back-up systems, such as a hand pump to replace the windmill pump, and buckets to replace the hand pump, have ensured continuous operation of the plant.

PROJECT COSTS

The overall original cost of the project was approximately $17,000, of which $3,000 was locally contributed in the form of labour, transportation, storage and working tools.

ANALYSIS OF THE FACTORS AFFECTING THE CONTINUED OPERATION OF THE SOLAR STILLS

The solar still in Source Philippe has now been in operation for some ten years. It is important to assess the reasons as to why this installation has been in continuous operation during this period and has become fully integrated into the life of the village community. It would certainly be presumptous on everyone's part to hastily analyse the factors and attribute the reasons to success to any one single component. The reality of most situations dictates that there are a variety of interacting factors and events which determine the reason for the success of this, or any other operation, for that manner. Unfortunately an indepth and rigourous academic study of this nature has not been undertaken. This would have required an interdisciplinary study of the village situation before the installation of the still, which would have as well analysed all the relevant factors affecting the socio-economic development within the community. As this has not been done prior to the installation and has only be undertaken marginally over the last decade, it suffices to say that any conclusions that can be drawn are speculative at best. Nonetheless sufficient experience is available so that a reasonable attempt can be made at this analysis.

In general when considering an Appropriate Technology Process, in its simplest form, one can use the following model to show the interaction between the various parameters listed below.

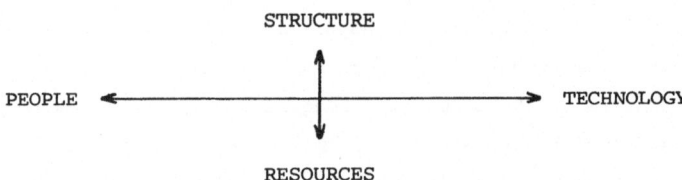

In the Haiti case the resources exist in the form of adequate saline water (and a severe lack of accessible fresh water), sunshine and a favorable climate as well as locally available materials. To this end a list of the groups involved in the project is included as well as the tasks that each undertook during the construction and later operational phase. The technology is appropriate. If it had not been so, the systems would have long fallen into disrepair and disuse due to technical difficulties. This is not to say that there has been no changes instituted. All Appropriate Technologies must be flexible and must adapt weaker components to improvements and modifications. This has been done in Haiti on a continual basis. Nonetheless the plant has continually been in operation for this period. Finally there is the population. The stills have replaced the well as the focal point for water collection, and the usual social activities associated with this type of operation. The population has

20

participated fully in the operation, maintenance and decision making regarding
the solar stills. What then is left? It is the structure and the structural
arrangements that have permitted the still to be in continous operation since
its inception. It is not possible that a more rigorous analysis would indicate
that the existence of an adequate structure has been one of the prime factors
which has enabled the unit to thrive and continue in its operation? Is it not
too often the case in developing areas that one either spends too much effort
on the socio-cultural aspects of the technology, or process or on the technical details to the detriment of the structural arrangements which will ensure
the continuity of the system? Technologies emanating from industrialized areas
may often be blinded to this reality in that structural and institutional arrangements are generally but not often stronger in these areas. It is not always the case however, for in Canada which is supposedly an industrialized
area, we often suffer, particularily in our outlying areas from poorer structural and institutional arrangements. As a result of this, not only are inappropriate technologies introduced, because the structural organization may
be thousand of kilometers away from the point of usage but often the very
structure which should be in place either does not exist or is not adequately
constituted. This probably occurs less in the smaller, more closely knit society of Western Europe and Japan or in the developed, institutionally oriented society in the United States of America. In the latter case, structures
are usually in place which encourage and foster development of all sorts. In
developing areas these structural arrangements exist primarily in the urban
and industrialized sectors. They are often weaker in the rural areas and in
this case resemble more closely the situation that have been noted in Canada
as discussed above. Is it not in this direction that we should concentrate
somewhat more of our efforts?

In the Haiti still case, the Eglise Methodiste has an adequate structural um-
brella with not only a local structure and village council at the site of the
still, but also an adequate support base in the capital city of Port-au-Prince.
While much of this is perhaps speculative, would not some of this account for
the continuing success of this installation? This is by no means to downgrade
the importance of the other factors and particularily the resolve and the
dedication of the local populations. But then, as has been cited earlier, is
Appropriate Technology not a harmoniuous interaction of many factors each of
with contributes to the overall success of a realization? It would be well
that we addressed more attention to these analyses if we are to find the gene-

ralized formulas and methodologies that will lead to an increased success rate of A.T. operations in the field. If this is to be done, we may have to structurally re-organize development projects and development aid to more closely reflect the realities of the situation in the field. This can only happen if there is a genuine committment and dedication to achieving the goal of the development of the lesser fortunate areas of the world. It there would be more committment of this nature, following overall appropriate methodologies, it would significantly enhance the use of appropriate technology processes in the field.

GROUPS INVOLVED IN SOLAR DISTILLATION PLANT

Group	Tasks
Eglise methodiste d'Haiti	- central coordination
	- management of work
	- transportation
	- social animation
	- tools
Government of Haiti	- permission for project
	- specialized testing
	- duty free status
	- some transport
	- testing in labs.
Haitian specialists	- technicians
	- workshop fabrication
	- social animation etc.
Brace research institute	- design and specifications
	- specialized technical labour
	- ordering materials
	- evaluation
	- site coordination
	- finance
Local population	- labour
	- some local materials
Oxfam-Canada	- finance
Overall	- much dedication

STATEMENTS

- For a long time people working in this field did not realize at all that they were working at appropriate technology. We try to satisfy a demand by examining various possibilities. We are not looking for an application for a piece of equipment we developed but we look for a demand and try to find out how this can be satisfied in such a way that the solution is fully integrated in the social system concerned.

- One has to realize that, however the people you work with may not be technicians, these people can make innovations, do suggestions and influence technological development. Technology and the choice of equipment has to be appropriate but may not be more than one-third of the problem. The rest of it be preparation, involvement and the structure encompassing everything.

- One has to have a good interplay between recources, technology, people and infrastructure. If that is not the case you will just be introducing a physical thing.

A NEW APPROACH TOWARDS RURAL DEVELOPMENT: APPROPRIATE TECHNOLOGY AND THE
EARTH-QUAKE OF 1976 IN GUATEMALA

*Roberto Caceres**

NEED FOR A NEW APPROACH

The conditions imposed by the earthquake which took place on February 4th,
1976 in Guatemala, marked a new process in the technological search. This
earthquake emphasized to two things:

1. A complete lack of profound and systematic knowledge of technical groups
   about the most crucial problems caused by the earthquake, such as:

   - lack of knowledge of a systematized technology for the use of *adobe* and
     earth in construction, in conditions of elevated seismic risk.
   - the relatively little advance of research and development of adequate
     materials for rural and suburban areas.
   - the lack of means and methods for a massive formation of rural manpower
     (promoters)
   - the existent breach between professionals with an academic formation in-
     adequate to deal with rural needs according to the rural and suburban
     communities feelings.
   - the lack of local ways of communication and groups participation.

2. The basic needs stated by the population, determined the importance of a
   new approach in the tackling of the same. The new approach should take into
   consideration this new situation which in *grosso modo* could be explained in
   the following way:

   - notwithstanding the deficiencies i: che formation of base groups there
     existed enough to start a new stage.
   - the willingness of the affected and non-affected population to collabo-
     rate in community development projects.

* *C.E.M.A.T., Guatemala City, Guatemala*

- the programs should be relatively unexpensive and should use a large
  amount of local resources.
- massive affluence of technical and financial aid which should be taken
  advantage of for the take-off of integral development projects.

3. From this permanent contact and emergency situation, maybe the most important criteria for the future development of affected areas are the following:

   - the importance of simplicity and humility in the work. This essential
     feature is characteristic for the new science and technology of the
     Third World.
   - the importance of the people's participation in the research, experimen-
     tation and development of alternative processes. In this case, we are
     trying to overcome a population-object situation in the investigation
     to a population-actor in the research.
   - the decisive emphasis that education should have towards an adequate ap-
     proach to motivate the population, form groups, evaluate results and by
     the follow-up guaranteed the input of the original.

ACTIVITIES OF RESEARCH AND DEVELOPMENT OF APPROPRIATE TECHNOLOGY REALIZED
BY CEMAT

I. Production and development of the "Cematita" Technology:

A. *Antecedents*

   After the 1976 earthquake a research was started for the investigation
   of intermediate materials of major resistance than the traditional
   adobe used in rural Guatemala. These materials should contain a small
   amount of cement, which is daily becoming more expensive and scarce because of the demand caused by the post-earthquake reconstruction. The
   need to investigate this alternative material was perceived through
   three types of explorations:

   1. Through workshops on popular housing in seismic zones, organized
      by CEMAT immediately after the earthquake.

   2. Interviews with national promoters who had knowledge of the diffe-
      rent potentialities and obstacles of production systems of construc-
      tion materials in Guatemala.

3. Through a documental study realized by the International Network on Appropriate Technology (RITA), a series of technological packages were found that we reused in other countries of the Third World and which are included among appropriate technologies.

The workshops on popular housing permitted the detection of a series of necessities and possible solutions. Specifically it is intended to emphasize the following:

1. The majority of rural masons and builders traditionally utilized lime in the preparation of mortars. Lime fundamentally exists at the north of the Motagua line in Guatemala. The mayas used lime in the construction of some fortifications such as Mixco Viejo. This archaeological verification helped us find a zone, within the affected area where lime was used a lot for construction. Actually, the rural people produce lime in very simple ovens, without elevation or chimneys, increasing considerate the amount of wood used per ton of lime.

2. The majority of participants to the workshops were inclined to massively use blocks of cement, due to the fact that it is much more resistant than adobe. This was demonstrated in those areas where only houses made of block were left standing. But at the same time, these rural masons didn't realize that the use of block would necessarily raise construction costs of a minimum rural house.

3. And last, it was established that in the affected area there was an abundance of volcanic ash which old masons call "selecto" and which is widely used to improve lime mortars.

With these conclusions, the existence of mortar which uses little cement and when compacted would be more resistant than adobe was investigated.

We were able to obtain information on the utilization of volcanic puzzolanic materials for the improvement of cement and the substitution and/ or complementation of the same by studying documents and the realization of trips in search of specialized information.

In some countries of the Third World, this type of material has been used as an alternative for rural infrastructures such as in the cases of Mexico, Peru, Chile, Ruanda, Tanzania, India, Germany, France and

26

Italy.

B. *Activities developed*

1. Exploration of mines and selection of a place for a pilot plant. It
   was necessary to search for this pyroplastic puzzolan which is abundant throughout the Guatemalan Highlands and lime which can only be
   found in pockets south of the Motagua River, in a relatively small
   zone. Based on the aforementioned, the existence of lime in San José
   Poaquil (department of Chimaltenango) was a determining factor in
   choosing this area, as the headquarters for the first pilot plant of
   Cematita.

2. Factibility and financial study. Since San José Poaquil is in the
   center of the area devastated by the earthquake, there is a large
   market for construction materials. Costs of material extraction, hand
   labor, infrastructure and administration were analyzed, thus obtaining a competitive price with other construction materials which are
   offered in the area.

3. Mechanical and chemical wheels were made to study and guarantee the
   mechanical resistance of Cematita in the Laboratory CEMAT has in
   Guatemala City and in the Laboratory of the Faculty of Engineering
   of the University of San Carlos in Guatemala.

4. Construction of the plant. CEMAT has maintained close work relationship with the Local Reconstruction Committee in San José Poaquil who
   has lent us a piece of land of 5.000 square meters in the center of
   the town, where we made a building which harbors the storehouses for
   prime material, the presses, the oven for the curing and the storehouses for the finished product. Also built was a house containing a
   kitchen and services for the workers.

5. Machinery and processes. The prime material (lime and pyroplastic
   material, Cematita) is kept in a dry place to be grounded and mixed
   afterwards in the hammer mill.

   The material is pressed and molded with a CINVA-RAM machine. Afterwards, the blocks are cured by vapor in the oven and left to dry
   during one week and are then distributed to the consumer.

   Other than the fabrication of blocks, Cematita has been used to con-

27

struct buildings using the "tapial" process (the material is pressed within metal molds which were designed by Mr. Bernardo Jelkman, collaborator of CEMAT).

6. Diffusion of Cematita and training of technicians. Several houses have been built which are used as prototypes for the demonstration of the material. A publication has been printed called "Cematita Project, An Alternative to Cement for the Rural Area", which will help in the realization of seminars with masons who are responsible for constructions in the communities of the region. These seminars will take place in San José Poaquil so as to learn the process practically. Groups of constructors have been sent to receive courses on "tapial" techniques.

II. EXPERIMENTATION AND DEVELOPMENT OF LATRINES PRODUCERS OF FERTILIZER

A. *Selection of an appropriate latrine*

The obviously deficient sanitary conditions of the Guatemalan rural area and the rapid contamination and environmental deterioration, has been clearly observed. The two main identified problems are faecal contamination, due to the lack of latrines or any sanitary facilities and, the second, the rapid deforestation due to the widespread use of wood for cooking in open ovens.

It is assumed that about 60% of the infantile morbidity is related to faecal transmitted infections, and that the average villager expends 20% of his working capacity in obtaining fire wood.

CEMAT's first line of action was the experimentation and development of alternative construction materials (cfr. Cematita) which would cheapen the cost of small rural infrastructures.

The second line of action was to develop alternatives for sanitary and cooling infrastructures. For the first we decided on three different prototypes of latrines for family and/or collective use and for the second, we chose the "poyo de lorena" (mud and sand stoves which economize wood). We will describe the "poyo de lorena" later on.

The first stage in the development of alternatives for sanitary infrastructures was the diagnostic of the most frequent causes for which the traditional latrines had not been widely diffused, even though there several programs during different times.

The initial conclusions after this preliminary exploration are the following:

1. The space occupied by the latrines is not productive and demands work, hours and money which does not visibly nor directly improve the family's income (the No. 1 priority of the majority).

2. The latrine, even though it is an intermediate technique between the toilet and defaecation on the ground, thus closer to rural conditions, enters into conflict with a cultural patron which has not been studied profoundly: defaecation on small pieces of cultivated land near the house (corn fields, coffee plantations, etc.).

   This tradition which has not been well studied, dates far back in the traditional rural history. We consider this as a traditional technique of *excreta re-use*, which is practiced in a more advanced way and more systematically in densely populated regions of Asia (Japan, China, Korea, Vietnam, and India).

3. The non-continuous characteristic of the type of latrines that have been diffused, means that the inversion is short-term which increases costs.

4. The non-existence of permanent educational programs adequately designed to motivate communities to make their own decisions and the necessary organization to carry out these programs.

With these elements which we picked up through indirect investigations, surveys, workshops and interviews with studied persons, we were able to determine the type of latrines which would be convenient to develop.

The following stage consisted in finding documentation on the existent experiences in the field of latrines with agricultural produce as a by-product. To provide a start for the exchange of information, the First National Seminar on Appropriate Technology was held in Panajachel, in 1977.

Once the national experiences were gathered, an investigation through the Appropriate Technology Network (RITA) was started concerning the most interesting experiences in this field, in different parts of the world. It's interesting to note that data banks on conventional technology did not contain any worthwhile information.

29

B. The anaerobic latrine for the production of fertilizer and bio-gas

During contacts and discussions with groups working with appropriate technology pertaining to the RENET Network, different models of continuous latrines were studied.

After studying the situation we decided to build a prototype which consists of a digestion chamber with an entrance and exit built into the same structure. It is a parallelepiped of 2.6 by 1.1 m. This box was built with local materials: volcanic rock, a poor concrete mixture (cement, lime and volcanic puzzolanic sand) and round stone. A slab of concrete was also cast using iron. The gasometers were built of barrels and the house of the latrines was made of wood.

For pedagogical goals and for experimentation purposes, a small digestor made of barrels was built which was filled with animal and agricultural wasteds. After approximately one month it began to produce bio-gas. This was very useful to convince the Local Committee. One of CEMAT's collaborators is a mason who decided to help us build a prototype. This has awakened a lot of interest in the communities around Lake Atitlán.

We learned that the Vietnamese during the war, had developed a double chamber latrine which occupies little space and which is hermetically closed after 1 or 2 months of use. One of the chambers is used alternately while the excrements compost is produced in the other chamber. Urine is evacuated separately. So far we have determined several advantages:

1. There is no contamination of subterraneous water

2. It produces fertilizer quickly

3. It produces less bad odor than the aerobic latrines

4. It occupies little space

5. It is cheap.

III. DEVELOPMENT AND DIFFUSION OF MUD AND SAND STOVES

A. Selection of the technology

Since the creation of CEMAT, there has been interest to learn and diffuse cheap techniques which would help to economize wood in the rural homes. Some 80% of the rural population uses wood as the only energy

source. We considered most convenient the diffusion of mud and sand stoves because of the following advantages:

1. The material of which it is made (mud and sand) are local

2. These materials are cheap

3. It economizes more wood than the traditional stove

4. The total cost, if made by the user is approximately Q. 20.00

5. To build a stove is easy to learn because the technique is relatively simple

6. It avoids the contamination of food while cooking

Another important feature is that the mud and sand stove designed by
ICADA CHOQUI, a group working in appropriate technology in Quezaltenan-
go, Guatemala, rapidly becomes hot since it is made of mud, which con-
serves heat during many hours, thanks to the system of tunnels which it
has, thus there is always hot water.

After a preliminary cost-benefit analysis we reached the conclusion
that the mud and sand stove could be a definite support to the rural
family's economy and also in the conservation of natural resources.
Anyway, it will be the families and rural communities which will decide
finally if it is convenient to introduce this type of technology.

Thanks to the cooperative relations existent between CEMAT and ICADA
CHOQUI, (Highland Experimentation Center), it was possible to train an
instructor of CEMAT in the construction technique of mud and sand stoves. This collaboration is part of the efforts of the National Network
on Appropriate Technology. So far, CEMAT has a team of rural instruc-
tors on the building of new stoves with different base groups around
several regions of the country.

B.   *The Poyo de Lorena*

The mud technology is very well known here, since the people are accustomed to mixtures of this material for construction purposes. These
stoves do not need special tools; only a shove, a machete and a kitchen
spoon are required for their construction; these are three tools that
even the poorest households have.

First, a mud base is built and then over the base, a solid block of wet mud and sand of about 40 cms. in height. When it is sufficiently dry, channels and holes for the pots are dug. To prevent the escape of heat and smoke, the pots have to be the exact size of the holes, to assure this, the pots which are going to be used are the ones used to take measurements from. The channel system is long and twisted to assure the absortion of heat before the smoke goes out through the chimney. Above the channels there are a series of temperature holes which diminish in size and which end in a fixed water heater, made of a 5 gallon can sunk into the block. It works as follows :

1. The stoves are monolitic therefore very strong

2. They are large, therefore they have a high thermic mass and retain heat well

3. Because they are principally of sand they do not crack, which is a very common problem with other stoves made of mud.

The heat of the fire box, heats the first pot and is absorbed by the walls of the box. From there on, the smoke goes throughout the channel system until it reaches the chimney, heating other pots and being absorbed. Gradually, the whole stove is heated. In the meanwhile, the smoke goes out through the chimney, leaving it's heat within the stove. We calculate that a well built model can retain 90% of the heat. People know that the stove cannot be touched once there is fire in it.

The heat is controlled by a series of regulators made of tin which open and close to control the air currents. With the regulators closed, the stove will heat the food during the night, cooking the black beans for breakfast while the family sleeps. Traditionally, the woman of the house gets up before the sun rises to start a fire to heat the coffee for breakfast. With this energy conservation system, there is hot coffee in the morning without adding new wood, showing great economy and comfort.

IV.   TRAINING OF RURAL PROMOTERS WITH APPROPRIATE TECHNOLOGIES FOR HEALTH

The lack of adequate approach of the Official Western medical system to attend the primary diseases of the rural Third World is obvious, as demonstrated in international meetings, to tackle such problems. For this reason, we have started a search of those technologies that could be economically, culturally and socially appropriate to the Guatemalan rural area.

These appropriate technologies for health include the validation of
Traditional Medicine; the systematization in the use of medicinal
plants, the transfer of appropriate technologies (such as acupuncture,
massage, chiro-practise) and, the improvement of the environmental quality by the appropriate technologies mentioned above.

For the diffusion of such technologies we have experimented on educational systems to transfer knowledge to the "Rural Health Promoter".
This training is done through workshops and follow-up activities during
a two year period.

So far we have been able to organize workshops in 4 different regions
of the country, with an average assistance of 25 rural health promoters, by region. With these promoters a network is in the process of
being formed which will provide primary health attention in an economi-
cal way highly acceptable by the traditional rural people. The market
for rural enterprises that commercialize such products is being studied
to promote the use of such alternatives.

- Choice of technology must always be a direct consequence of participation of the people. The best way to find solutions in this field
  is to work on the family level.

- To have a spreading effect on the regional level it is necessary to
  organize the population in some way. This second level of organisation should be small entreprises coordinated on a regional level.

- Most important obstacles in development come from technicians on the
  national level.
  Their education is based on a very sophisticated concept and stands
  far off the people.
  A main task in thus to bring changes in their education.

DIAN DESSA

*Anton Soedjarwo**

We work in the area of Dian Dessa since 1966. In the beginning of the project
we had no specific approach to this type of work and we were not familiar with
concepts such as basic needs, appropriate technology, etc.. May be we work in
these rural areas just by chance. We just went there; we found out about reality and saw that this was very much different from the things we discussed in
the University.

Today we have several sections such as watersupply, energy (e.g. waterturbine,
windturbine, bio-gaz, etc..), agriculture, animal husbandry, food technics and
ferro-cement. We have also a social science group and a traditional technology
section. Especially the traditional technology section is interesting. We
think that even when traditional technology can not answer questions as posed
to us now, it can learn us about the perception of these things and for instance shows us the perception of local recources as used by the local community. If we learn to understand about this that will help us when introducing
a new technology or modifying a traditional technology. This is important because in some cases we found that even a small modification of a traditional
technology meant the difference between life and death for thousands of people.
I will explain this with a few experiences we had in projects for water supply.

First is about the introduction of a rainwatercollector in the East District
of Java Island. First thing to realize is that the social structure in various
parts of the country is completely different from one part to another. This
is sometimes even the case in one district. Traditionally the people in the
South of Java took their water from "lakes". This all happened in a very stable system which was under supervision of a Sultan. This Sultan was also the

**Yogyakarta, Indonesia*

owner of all the land. In 1942 the Japanese came in. Theyoccupied the land,
cut down the trees and the whole eco-system around the lakes became unstable
and a shortage for drinking water came into existance. We found out that
following the perception of the people, they think that water is something
coming from the sky. In a big, O.D.M. sponsored, project the digging of deep
wells was proposed and up till now none of the wells is used by the people.
This may be so because the use of wells is not in accordance to their traditional way of thinking. We therefore started with the idea of the rainwatercollector. We started to make only a few examples, using local materials. We
built various models, using different materials (e.g. bamboo-cement and
ferro-cement) and let the people choose. Up till now in that area already 1840
tanks have been constructed and these will serve approximately 11.000 people.
We see that this process is still going on.

We also tried the same approach at Madura and a few other small islands, but
found a completely different situation there. I speak now of a small island
with a length of 2 kilometers and a width of approximately 700 meters. It has
a total population of 12.800. U.N.I.C.E.F. worked in that area untill 8 months
ago. We found out that they worked as real donors. They gave rain water collectors but only the rich people could afford these and started to sell water
to the poor. This hence raised a completely different problem. You have to
realise that not to many people work in that area and social systems are very
traditional. The rates in the area are extremely high. This means that if
somebody gets a wage of 20.000 rupees this person has to work for this money
for ever. The worst thing is however that if this person dies, his son has to
replace him. In fact nobody is born free on that island. The only free labour
there are the women. For that reason we work with the women there because to
our opinion they are the only group who can possibly solve this water problem.

Another case we found at Madura Island.We wanted to start an environmental
program there. After three months of talk we became very frustrated. We talked
with them about erosion and all problems related to that issue but nobody understood the problem. We then realized that these people are very fanatic
moslems and we started to study the Koran. Among many words in this book we
also found something about water and environment. For example we found that
the word "Kalifat" means that everybody is born to take care of his surroun-
dings; e.g. wife, children, relatives but also the forests, etc.. I agree with
the statement that technology may just be one third of the problem. Social

infrastructure may be the most important element. It is true that this is a main element about engineering which up till now is not introduced in the Universities.

We see today that to many organisations act as supplyers of means. However it seems that sometimes local people start to participate this generally is not the case. In many areas you can find people working in what we call appropriate traditional technology. These people are hardly ever organised, nor can they deal with people of any government organisation. They would just not be able to fill in the forms because they can't write or read. May be these people only need a very small amount of money, just to raise their standard of living a little. To them the question about the main amount of money on the top level and how to touch this is not a relevant one. May be Schumacher was right when he stated that "Small is beautiful"; in many cases Small is very painful.

It is not easy to identify the limiting factors in these sort of processes. There may be many. Sometimes it be the funds and sometimes ability. Main thing is to get real motivated people who really want to do something in rural areas. Also in the case of Indonesia it is clear that there can not be an easy transfer of solutions from one place to another. Indonesia from Sumatra to Irian is a distance longer than from London to Istanbul. The case is to understand the perception of technology and than decide about the next steps to be taken. It must be realized however that in such processes always a different singer sings a different song.

STATEMENTS

- Even small modifications of traditional technologies can mean the difference between life and death for many people in rural areas.
  Studying traditional technology is a basic element in rural development.

- The most important element in development may be social infrastructure.
  This issue should be introduced also in engineering educational programs
  at universities.

- We have to realize the relativity of the utterance "small is beautiful";
  we must realize that small sometimes can be very painfull.

URBAN AGRICULTURE AS AN APPROPRIATE TECHNOLOGY

*Tom Fox**

With the emergence of hundreds of community gardens in urban areas across the
country during the early 1970's, people began to realize the potential of
food production in cities. Open space, once only a social amenity in neighbourhoods, began to take on new dimensions. Urban agriculture is still a relatively fragmented movement, which consists primarily of many small-scale
demonstrations. It is important, however, to keep in mind that a system is
developing. When we look at the integration of demonstrations we begin to see
possibilities for the years ahead.

FOOD PRODUCTION

Community gardening has taken a strong hold in many cities. Allotment gardens
and individual plots number in the tens of thousands nationwide. California
has a state Coordinator of Community Gardening, and urban gardens number 100
statewide. New York City alone has 2400 urban families producing 270 tons of
fresh produce annually, according to Cornell University estimates.

Community gardens are developing as self-reliant entities in many communities.
Neighbourhood compost piles recycle household organics. Municipal leaf dumps,
zoos and stabels are great resources for urban gardeners searching for organics. Winter rye and buckwheat are being grown as people learn the importance
of cover crops and green manure. Organic gardening practices are being applied
in an ever increasing number of urban gardens. In Boston, irrigation systems
from city water lines have been installed in some urban gardens. Cold frames
and grow holes are now used to extend the growing season and increase garden
productivity. Bee hives and beetle traps are being used in Hartford as people
realize the interrelationship of plants and insects.

**New York, United States of America*

Some cities, such as Davis, California, have virtually eliminated the use of toxic sprays by instituting integrated pest management programs.[1] The draft plan for the revitalization of the South Bronx, submitted by New York City to the Federal Government, included an urban farm.[2]

The ownership of land is taking new forms as pioneer community gardeners are pushed from their rented or donated land by rising real estate values. Community groups in Brooklyn's Bedford-Stuyvesant are buying vacant city owned land at public auctions. Land trusts are developing in many urban areas. Community groups have incorporated land trusts and are collectively managing community gardens in Newark, New Jersey.

Greenhouses are being tested in many urban areas. In a densely populated neighbourhood of Chicago, a rooftop greenhouse is producing crops year round. The greenhouse at the Institute for Local Self-Reliance is being monitored to see if a small solar structure can produce enough food for a family and how much time and energy is involved. Attached or lean-to greenhouses have been developed for less dense urban areas. In Cambridge, Massachusetts, a lean-to greenhouse uses greywater effluent from the household for nutrients and water. The top of the line in solar greenhouses, the Ark, is being perfected in Woods Hole, Massachusetts and Prince Edward Island, Canada.[3] A similar structure which contains aquaculture, vermiculture, horticulture, mushrooms, rabbits and poultry is a proposed development on E. 168th and Washington Avenue in the South Bronx.

With an increase in food production has come a need for preservation and storage. Community canning centers have been attempting to provide low cost food preservation for gardeners. Solar food dryers have allowed individuals to preserve food grown in community gardens with no need for fossil fuel supplements.

There are, however, limits being evidenced in urban food production. Some of these are inherent in the design of urban areas which were not planned or developed for farming. Others are due to the high amount of heavy metals found in the air, soil and water in our cities. People are beginning to see the scale of production necessary to feed our urban populations is not something that can be realized in cities.

Buying clubs and food cooperatives are giving consumers access to quality produce and a realization of their collective buying power. In Chicago, the Self
Help Action Center has helped start over 300 co-ops. They have 35 constituent
co-ops with approximately 7,860 household members, and they estimate co-op
members save 30¢ on every food dollar in winter and 50-60¢ in the summer. In
Seattle, the Bulk Commodities Exchange brings local farmers together with
restaurants, retailers, co-ops and buying clubs. Farmers averaged 18% higher
food prices and buying club members saved 40% over supermarket prices.[4]

Direct marketing has been flourishing in inner city areas around the country.
Evening and weekend farmers are becoming commonplace in the parking lots of
Hartford, Boston and New York. Farmers, some accepting food stamps, have been
dealing face-to-face with inner city residents. The communication and educational processes that occur at farmers markets would be difficult to dupli-
cate through any other medium. Greenmarkets in New York City were at 8 different locations last year.[5] They handled $ 1 million in produce.

As we look at alternate marketing, we are also looking at a new type of
small business development potential. Cottage industries, such as sprouts and
worm farming, are low in economic return and high labor intensive businesses,
but they have their niche in an urban agricultural system. Composting operations are testing the economic feasibility of capitalizing on large scale
municipal organic waste resources. Conventional technologies are being aug-
mented with alternate technologies to reduce operation costs. The Bronx
Frontier Development Corporation is building a 40 KW windmill to power a forced aeration system which will, in turn, increase their composting efficiency and at the same time provide all the electricity needed to power their
field office. Because high energy costs are having a disasterous impact on
small businesses, the use of alternate technologies is essential in the creation of sound small-scale business development in urban areas.

A recent suggestion has been to grow ornamental annuals and perennials on
vacant lots and do direct marketing in downtown business districts. The nursery industry that fled the cities has a chance to be reestablished now that
large amounts of open space, created by urban renewal, have cropped up in
most inner city neighbourhoods. The City of Jackson, Michigan has created

nine small nurseries for street trees on vacant city-owned land. If cities
instituted policies to purchase trees from urban nurseries it could insure the
city of getting pollution tolerant trees, keep tax dollars in the city, and
create neighbourhood businesses that are environmentally sound.

Whether we develop systems that sustain themselves or generate some profit,
alternate marketing systems are a major ingredient in urban agriculture.

HORTICULTURE AND ENVIRONMENTAL MAINTENANCE

Another aspect of increased environmental awareness has been the approach to
open space as it now exists in cities. People remember when the streets had
trees that shaded and cooled the neighbourhood. They remember when parks were
splashes of color and that vacant lots were green years before they became
depositories for broken bottles and discarded wastes.

In Los Angeles, 100.000 trees have been planted on city streets, yards and
parks by the Tree People. Most cities have reacted to the death of their
urban forests by paving over the tree pits so cars could park more easily and
maintenance costs were kept low. Philadelphia Green has been using a concrete
saw to make new tree pits ans has helped residents plant their own trees.
In New York City, the Street Tree Consortium offers a twelve hour course and
exam that will license community members to care for newly planted trees.
The Oakland Tree Task Force has assisted communities in planting 2.000 trees
around their city.

Community parks are being developed all over the country. In some, trees are
being selected for their fruit production, fragranceproduction, micro-climatic
impacts, and harvesting qualities as well as their pollution tolerance. There
are flower gardens of plants designed for picking and annuals in pots so
children can rearrange gardens without disturbing the vegetation. Plants
that attract wildlife, by providing food or shelter, vines that cover fences
and screen gardens from traffic exhausts, and ground cover that carpet shady
areas and curb erosion are all essential to the urban ecosystem. Window boxes
and rooftop gardens are bringing color and life to the concrete hills we live
in. Grassy playing fields now cover the 5½ acres of land that was New York
City's last asphalt plant. [6] *"Asphalt Green"*, as most of these other examples,
was a citizen based operation that fought against an unresponsive municipality

ignorant of the needs of its citizens. With positive municipal response to strong citizen activism, examples of the potential for liveable cities are appearing.

The technical assistance necessary for citizens to develop green spaces in cities has come both from traditional and non-traditional sources. The Brooklyn Botanic Garden has a community gardening specialist, a trained horticulturalist, who works with local community groups who need professional help. The Green Guerillas is a non-profit group of volunteers who have expertise in many aspects of urban agriculture. They provide free technical assistance to community groups and have a plant nursery where plants are given away to community gardeners. Some generous people in Westchester County tag the trees and shrubs that have self-seeded on their property. The Green Guerillas rent a truck, ball and burlap the plant material, and bring it to the urban nursery to be given away to community gardeners.

EDUCATION

One of the most positive benefits of the urban agriculture movement is that a wealth of information about farming practices is being shared. Many inner-city children have never seen a farm or had a garden, but their parents or grandparents have. They may come from small farms in the South or from rural populations in the Carribean, Central and South America. For many of these people, who have left their rural roots, it is an opportunity to get back to the land. They are planting cultural varieties of vegetables and ornamentals that are adding diversity to the system. Most important of all, they are teaching their children. Society has made them relatively unimportant as teachers they don't have standard educations, but they are now in a position of knowing something their children want to know. In Washington, D.C., local youths prepared garden plots for senior citizens and did all the heavy labor tasks. The seniors provided the technical assistance and maintenance and produce was shared.

School systems have been adapting curriculum and working with other government agencies to broaden the scope of education. In New York, 3.500 school children from inner city areas went on *"Operation Explore"*. This program took them on day long trips to farms and beaches to see where food comes from and explain food chains. Nutrition education follow-ups in school and trips to local community gardens helped to round out the experience. At Edward R.

Murrow High School in Brooklyn, agriculture students have learned to do ornamental planting and started a tree nursery at nearby Gateway National Recreation Area. They have raised plants in their greenhouse and distributed them
at hospitals. They have begun a campaign to green the local business district
with funds that are being provided by local merchants. College interns are beginning to get experiences in urban agriculture in places like the Institute.

Non-profits are making tremendous efforts in the education field. The *"Chuck-
wagon"* is a nutrition education mobile unit. It is a converted library book-
mobile that has a kitchen and classroom space. It tours South Bronx schools in
the spring, fall, and winter and community gardens in the summer, involving
both children and adults in cooking fresh foods and explaining nutritional
needs and cultural dietary differences. When the Trust for Public Land con-
ducts a workshop on land trusts, it teaches community residents about their
own ability to acquire and control land.[7] The community design workshops
offered by the Institute for Local Self-Reliance give people the knowledge
they need to make choices about the trade-offs necessary for integrated, open
space development. The goal is to transfer a knowledge base. A community cannot care for a garden, park, or nursery unless it has the tools to do the
work. With alternate educational opportunities in school, and in the neighbourhood, there is a chance that urban agriculture will become a community
skill.

INTEGRATION OF URBAN AND RURAL

Cities were not designed to grow food! They were planned and built to house
people and produce manufactured goods. Cities have seen a great exodus of
commercial business to their rural rings. Farms continue to be gobbled up as
corporations build energy intensive monoliths on some of America's most productive farmland. With food prices soaring, people trying to adapt cities to
become self-reliant food producing units are realizing it is just not possible. Cities can produce food, but the majority of it must come from the surrounding region.

The key to any future agriculture system is to consider production and resource distribution on a regional level. Direct marketing is bringing farmers to
the city and beginning to prove the economics of small scale truck farming.
A two-year, $ 5 million pilot program in Massachusetts is assessing the fea-

44

sibility of using state funds to acquire the development rights on rural
farms. Although this program is very limited, it has a potential for aiding
in the preservation of our country's dwindling number of family farms. These
programs coupled with direct marketing systems may help to reestablish human
scale agriculture on a regional level.

The first study on the economic feasibility of regional composting has been
done for Omaha, Nebraska.[8] The study looked at all the existing sources of
organics in a three county metropolitan region including sludge, wastes from
approximately 20 packing plants, and race track and stockyard manure. They
then developed different planting schemes to utilize these wastes. In a six-
year cropping sequences of corn, soybean, and alfalfa it was estimated that
by 1994, the available organics could supply the fertilizer needs for 80.000
acres for five of the six years. The sixth year figures fell short in nitro-
gen available but were sufficient in phosphorus and potassium. The study did
not include household wastes, which the researcher admits should be looked
into, and there was still a substantial agronomic benefit to regional composting.

PROBLEMS AND QUESTIONS FOR THE 80's

One of the major problems facing urban agriculture is the modern design of
cities. The current method of redevelopment in our inner cities is high-
density and centralized, presenting a tremendous stress on living systems.
The diversion of rain water to sewers along concrete and asphalt streets and
sidewalks is starving the environment. Most of our new development is geared
for low maintenance and machine compatibility. We will never be able to develop viable urban areas unless future design is geared towards maximum rehabilitation, low density new construction, and the redevelopment of environmentally sound systems. Management techniques will have to reflect this
change to curb the loss of already existing resources like our urban forests.

We will have to move towards a more labor intensive agriculture system. Urban
gardens, nurseries, family farms and alternate marketing systems all require
people power. Energy cost effectiveness will be increased if human energy
lessens our need for fossil fuel energy. The change from a capital intensive
agriculture system must be made if we are to overcome the problems of malnutrition, high unemployment and increasing energy costs. Making maximum use of

available human potential will be the only way to make a system work.

As more and more data is compiled on the effects of heavy metals the major question has become *"What food should we grow in our cities?"* Automobiles, lead based paints and industrial pollution have inundated our soils with lead and cadmium. A recent Cornell University study recommends that cabbage and other heavy metals accumulators not be grown in urban gardens.[9] They may increase the already high amount of these substances ingested by urban dwellers via the air and water. But other studies in New York have shown that store bought produce contains similar levels of heavy metals and suburban soils tested in Boston have also shown elevated levels of these substances. Heavy metals are now ubiquitous substances and how this problem is dealt with will be important for both urban and rural agriculture.

The Boston Urban Gardeners have worked closely with the U.S. Environmental Protection Agency and the Massachusetts Department of Food and Agriculture to establish a heavy metals testing program. Five hundred plots from 65 urban gardens have been tested and parameters for growing produce are beginning to be defined. Advertisements in the Boston Globe have greatly increased participation and gardeners in dangerous areas are being advised to relocate. If we couple these programs with blood testing programs, especially for inner city youth, we may be able to minimize the problem. We must find a way to reduce inputs of toxic substances to our food chain. This means extensive studies on applications of sludge to agricultural lands before the practice becomes uncontainable.

Better food storage systems in and around cities will have to be established to reduce the need for processing and packaging, another area of both contamination and waste. Increasing food stamp allocations to complement regional production cycles would give the poor greater buying power. The proper design and distribution capability of this system should reduce the amount of transportation currently needed. The focus should be on utilization of already existing mass transit systems to reduce the impact of new construction. Rebuilding the nations railroads to the farms should be a priority.

The ability of the federal government to respond to present initiatives will be crucial. The U.S. Treasury is the only tool within the citizens power that is large enough to stimulate change in the system by providing research and development monies. The federal government is in a position to coordinate interagency information transfer and develop national policy. The primary purpose should be to stimulate small scale economic development as they develop national policy.

The cornerstone for the future is demonstration projects. When they succeed or fail we have to know why. All available data must be collected, stored, analyzed and made available to the public. By documenting demonstrations and highlighting their strengths and weaknesses - a relevant urban agriculture system can be developed. If we rush to enact policy that is not timed with the completion of successful models we will continue to suffer under our present planning malady. Good paper studies do not always make good real word programs.

BIBLIOGRAPHY

1   *Olkowski, W. and H.   Developing Urban IPM Delivery Systems. Paper delivered at IPM Conference "New Frontiers in Pest Management," Sacramento, CA, 1977.*

2   *New York City Planning Commission, The South Bronx, A Plan for Revitalization, December 1977.*

3   *Edited by Nancy Jack Todd. Book of the New Alchemists, EP Dutton, NYC, NY, 1977.*

4   *New Directions in Farm Land and Food Policy: A Time for State and Local Action, Conference on State and Local Public Policy, Washington, D.C., January, 1979.*

5   *Benepe, Barry. Greenmarkets "The Rebirth of Farmers Markets in New York City". Council on Environment, New York City, 1977.*

6   *New York Times article - "Asphalt Green.", April, 1979.*

7   *Citizens Action Manual, Trust for Public Land, San Francisco, CA, February 1979.*

8   *Blobaum, Roger, et al. A Potential for Applying Urban Wastes to Agricultural Land, Roger Blobaum and Associates, West Des Moines, IA, January 1979.*

9   *Proceedings Toxic Elements Studies - Food Crops and Urban Vegetable Gardens sponsored by Cornell University Cooperative Extension at Wave Hill Environmental Center, Bronx, New York, June 14, 1978.*

RESOURCE ORGANIZATIONS

1   California State Community Gardens Office   Rosemarry Menninger
    Office of Appropriate Technology
    1530 10th Street
    Sacramento, California 95814

2   Cornell University Cooperative Extension   Sam Segal
    11 Park Place
    Room 1016
    New York, New York   1007

3   Knox Foundation   Mike Marcetti
    One Constitution Plaza
    Hartford, Connecticut 06103

4   Magnolia Tree Earth Center   Joan Edwards
    1512 Futton Street
    Brooklyn, New York 10112

5   Trust For Public Land   Peter Stein
    95 Madison Ave
    New York, New York 10016

6   Center For Neighbourhood Technology   Scott Bernstein
    570 West Randolph Street
    Chicago, Illinois 60606

7   New Alchemey Institute   John Todd
    Woods Hole, Massachusetts 02543

8   People's Development Center   Ramon Rueda
    500 East 16th Street
    Bronx, New York 10456

9   Self-Help Action Center   Dorothy Shavers
    5938 Ashland
    Chicago, Illinois 60636

10	Hunger Action Center	Cindy Solie
	Alaska Building	
	Room 300	
	Seattle, Washington 98104	

11	Greenmarkets	Barry Benepe
	24 West 40th Street	
	New York, New York	

12	Bronx Frontier Development Corporation	Jack Flanagan
	1000 Leggett Ave	
	Bronx, New York 10474	

13	The Tree People	Andy Lipkiss
	California Conservaiton Project	
	12601 Mulholland	
	Beverly Hills, California	

14	Philadelphia Green	Blaine Bonham
	325 Walnut Street	
	Philadelphia, Pennsylvania 19106	

15	Street Tree Consortium	Liz Christy
	c:o Council on the Environment	
	51 Chambers Street	
	New York, New York 10007	

16	Oakland Tree Task Force	Dana Cole
	1419 Broadway	
	Oakland, California 94612	

17	South Bronx Open Space Task Force	Amos Taylor
	740 Kelly Street	
	Bronx, New York 10455	

18	Brooklyn Botanical Gardens	Lucy Chamberlain
	1000 Washington Ave	
	Brooklyn, New York 10003	

19   Green Guerillas   Tim Stinhoff
     P.O. Box 681
     Cooper Station
     New York, New York 10003

20   Gateway National Recreation Area   Sam Holmes
     H.Q. Building 69
     Floyd Bennett Field   .
     Balyn, New York 11234

21   Suffolk County Food & Urban Gardening   Eileen Mc Donough
     Urban Gardening Program
     University of Massachusetts
     Downtown Center
     Boston, Massachusetts 02125

22   Blobaum Associates   Roger Blobaum
     1346 Connecticut Ave NW
     Washington, D.C. 20036

23   Boston Urban Gardeners   Judy Wagner
     66 Hereford Street
     Boston, Massachusetts 02115

24   U.S. Environmental Protection Agency   Tom Spittler
     60 Westview Street
     Lexington, Massachusetts 02173

STATEMENTS

There is an overall important task for the educational system. Too long people
have been studying only the mechanism of production and consumption. There is
a need for a long term process that will take people back to the situation
that they will learn to interprete their environmental situation as a critical variable in system development.

In industrialized countries appropriate technology means different things to
different people. To the rich it is a way of reducing their costs, to the
poor it is a way to survive and live a better life. Smaller scale technologies
are beginning to be viewed as something necessary for everyones well being.
But its impact has different degrees of magnitude depending on need.

We must look at appropriate technology as something that is relevant to the
rich and poor, developed and developing, urban and rural. If the rich don't
reduce their consumption there will be little resources for the poor to work
with. If we concentrate our efforts on rural populations we will miss the
mark. In most developed countries the majority of the population lives in
and around cities. In many developing countries there continues to be a population shift towards the cities.

Appropriate technology applications must not be limited to rural populations
in developing countries. These technologies must have the broadest possible
applications to have an impact on the scale of the problems we now face.

Session 2: Organization of Appropriate Technology Activities

CONCEPTS AND MODELS FOR THE DEVELOPMENT OF APPROPRIATE TECHNOLOGY FOR
RURAL AREAS

*Mansur M. Hoda**

INTRODUCTION

Technology involves the application of science and knowledge to practical use, enabling man to live more comfortably and securely. Fire, the
wheel, the tools use in the stone, bronze and iron-ages and the industrial revolution are all links in the same chain. Man being a rational
animal, could manipulate, exploit and use various natural materials such
as minerals, plants, animals and energy for this purpose. After hunting
animals, man developed an agricultural society and later an industrial
civilisation as he followed the path of development and progress, all
the time using his knowledge, ingenuity and craftmanship to develop all
sorts of tools and contrivances. Occasionally, he suffered a setback
in this process because of the long term adverse effects of a technological innovation - for example, the introduction of extensive and intensive agriculture and indiscriminate deforestation , mostly in Asian
nations, led to soil erosion and the creation of deserts and dust bowls.

The advent of the industrial revolution, which was brought about by the
application of mechanised power to production, gave a qualitative jump
to this process of civilisation. Prior to that man's work was generally
carried out using human and animal power with very little mechanical
energy derived from an external source being used. Fire was the only external source to be used for cooking food, heating, softening and melting
metals and minerals. As agricultural technology had brought human beings
out of the forests and concentrated them in organized settlements, so mechanised power, which could be applied on a much larger scale, created

A.T.D.A., U.P., India

a high concentration of productive activities, thus leading to the creation of large urban complexes, and the concentration of wealth, political and economic power in the hands of smaller groups.

In the Western industrialised nations, the rural societies dwindled and shrank to 5% to 6% of the total population.

The main advantage of mechanised power was that it concentrated production in one area, yielding larger profits because of easy management. However, in recent times these advantages have been offset by recurrent labour disputes which can also be easily organised in concentrated areas and result in loss of production and the exertion of pressure on industries. The organisation of industry became so complex that management started depending on new techniques of management, such as remote control, computerisation etc.

LARGE VERSUS SMALL

There are certain rules for the growth of industries and companies in the same way as there are rules for the growth of animals and plants. After some time, industry and companies like cities become unmanageable once they exceed a certain size. Efforts should therefore be made sufficiently in advance to determine the optimum size of an industrial activity. Dr. Schumacher invented the slogan "Small is beautiful" in defiance of the modern trend of 'the Bigger the Better'.

We cannot always blame modern science and technology for the overgrowth in the size of industrial activity. Some modern developments have actually helped in decentralising industries and establishing smaller size holdings. When steampower held away, engineering workshops used to be very large to take advantage of the large boiler and steam engine efficiencies necessary for providing motive power to the machines. However, the introduction of electric power reversed this trend and we can now see tiny cottage workshops at the side, even in rural and semi-urban areas, run by one or two people. Many of the engineering workshops are now back to the size of the ancient Smithy, although they use more up-to-date techniques.

It follows therefore that it is possible to use modern inventions to reduce the size of industrial holdings and to disperse them widely for the sake of employment generation and the balanced development of a country.

ENERGY AND POWER

The motive power for modern industrial development is almost totally based on
electricity, which can be easily transmitted by cable throughout the length
and breadth of a country. The main problem is how to generate electricity itself. Thus far, fossil fuels, viz., coal and oil, have been almost exclusively
used for electricity generation. But supplies of these will soon be exhausted.
There is a danger of industrial civilisation collapsing completely if no alternative method of generating electricity is found soon. A great effort has
been made in the last two decades to replace fossil fuels by nuclear power.
However, nuclear power has not yet won the confidence of the managers of the
modern world as a potential source of energy because of incalculable safety
problems and invisible radiation hazards. The main question therefore is :
- Whether it is possible to generate electricity or use mechanical energy
  based on income energy. The most important forms of income energy available
  at the moment appear to be the following - :

              1 Solar
              2 Wind
              3 Mini Hydrel
              4 Wood
              5 Animal dung

All these are available in abundance and will be automatically replenished as
they are consumed. Wood can be replaced by human beings through afforestation
and tree planting on a systematic basis.

LIFE STYLE AND SUITABLE PRODUCTS

We have to calculate whether all the above mentioned sources of energy will be
capable of providing sufficient power to sustain mankind in the extravagant
manner in which the people of the industrialised country have become accustomed to live, and the people of the developing countries are aspiring to
live. If not, what kind of life style can they sustain? Would it necessitate
a drastic reduction in the life style of all people throughout the world?

This brings us to the question of the selection of products that will find a
ready market and yet will not tie the people of the world to a life of luxury

and extravagance. A reappraisal has to be made of the priorities for newly-selected products, which is based more on need than on greed. During the past century in Indian situation we have seen an influx of new products such as man-made fibres displacing cotton and wool, whiteware pottery displacing red-clay pottery, fountainpen displacing reed pens, toothbrush replacing datwan (a tree twig used for brushing teeth) tractors displacing bullock driven ploughs and implements, cement and concrete displacing mud, thatch bamboo and timber, white crystal sugar displacing gur and khandsari, Dalda (hydrogenated vegetable oil) displacing vegetable oil or ghee, plastic replacing wood, metal, leather etc. etc. *

ADVERSE EFFECTS OF TECHNOLOGY

The development of technology is no longer a smooth and easy process. The erratic effects of modern technology are perceived in a very short time. Earlier when agricultural technology was introduced and forests were destroyed to set up a civilisation based on cultivation, the adverse effects were perceived in thousands of years. But now, after only 200 years, industrial civilisation is being threatened by a serious crisis. The principal threats to this civilisation are : -

1 Exhaustion of resources on an unprecedented scale;

2 Pollution of land, rivers, oceans and the air by effluents and wastes, fumes and radiation released by industries in huge quantities;

3 The breaking up of the social fabric due to methods of production which alienate the human being from the production process and make him into a machine tool or a cog in the wheel of production. This has resulted in a large number of dropouts in highly industrialised countries and vast income differences in the developing poor countries. Because of this and because of the very low level of production in the developing world, the thought was conceived of developing a technology more in tune with nature, more permanent and peaceful and giving a better income to the worker and artisans in the developing countries. This has been referred to by various names but the most popular usage is the 'appropriate technology'.

APPROPRIATE TECHNOLOGY

The concept of appropriate technology endeavours to eliminate the adverse ef-

*The same may apply with a little variation to other developing countries.

fects of modern technology by devising technologies for peace and permanence, by making units as small as possible, by dispersing them over a wider area and by ensuring a living wage to the people, who work at it. It also tries to change the life-style of the world, characterized by extravagence and conspicuous consumption by bringing mankind back to a life of simplicity which is in harmony with nature. Thus under the production programme of appropriate technology, a new list of products based on this ideology has to be prepared. One of the criteria for fixing priorities is the meeting of the essential needs of human beings.

APPROPRIATE TECHNOLOGY FOR AGRICULTURE

Food will naturally take pride of place in this list because no one can survive without food, other essential needs such as air and water being supplied free by nature. For producing food, an efficient and appropriate method of agriculture based on the criteria laid down above has to be developed. In developing countries this raises two basic questions. Firstly, whether sufficient return can be obtained by subsistence farmers who have to make an increased investment according to new technology of farming. Even if by so doing the productivity of the farm is increased which is obviously in the national and international interest, the interests of the individual farmer have also to be taken into account. If by investing £ 2/- a farmer can get a return of £ 4/- it is a 100% return on his investment. On the other hand, if by making an investment of £ 6:- he gets £9:- this is only a 50% return on his investment, although the output of his farm has increased from £ 4/- to £ 9/-, which will add to the total reserves of food.

INPUTS

The second basic question is whether the inputs of technology and material can be obtained from the rural area itself or will have to be brought in from cities and urban industries. In the latter case, the resources of the rural areas will be siphoned off into urban areas which the concept of appropriate technology tries to restrict. In the calculations given above, the total investment of £ 6/- was for fertilizer, irrigation, high-breed seeds and mechanised equipment which are usually obtained from urban areas. This siphons off the resources from rural areas to the urban area and further impoverishes the rural population. In the earlier case, the investment of £ 2/- in the form of local seed, green and organic manures, bullocks and ploughs and human labour will all be retained in the rural area itself. It should therefore be borne in mind that unless improved agricultural implements or new techniques bring about a tremendous improvement in productivity and quality of the product they should not be recommended. If a small tractor ploughs the same area of land in a day as the bullocks there would seem to be no sense in introducing tractors which consume large quantities of oil, purchased from urban centres.

Similarly, chemical fertilizer brought from factories in the urban centres do not make a great deal of difference to production in the long run, while they do adversely affect the fertility of the soil. Would it not be better to use organic manures like animal dung, properly composted, instead. Animal dung is now used extensively as kitchen fuel and we will refer to this later. As far as green manures are concerned the problem for the farmer is that it keeps the field engaged without giving him any cash return from the green manure crop. If some variety of green manure such as Indigo could be introduced which would give a cash return to the farmer in some form they would be able to raise the crop of green manure for the benefit of the land. This would be much more useful than chemical fertilizer.

The main problem is that the *new techniques and processes introduced in agriculture* have so far been taken advantage of by large and wealthy farmers who own vast tracts of land to which the new techniques are best suited. In addition to buying their inputs from the urban areas these well-off farmers become accustomed to luxuries and spend their surplus income on conspicuous consumption and exotic goods from the cities. This further siphons off the wealth of the country-side for useless goods, such as cars, refrigerators, terylene and polyester clothing, plastic goods, watches etc.

The important areas in agricultural technology are *irrigation facilities,* harvest and post-harvest and processing techniques and storage. The most widely used irrigation method at present is by tubewells. This again helps mostly the wealthy farmers, lowering the water level, drying up wells and tanks and the poor farmers are further deprived of water facilities. In the areas which are irrigated by canals, the situation is a bit better. However, the poor farmer has to depend largely on rain water and thus his choice of crop is strictly limited.

For *harvesting,* human labour is still used on a very extensive scale, creating a shortage of labour at harvest time. There is perhaps a need for an appropriate implement for harvesting which can be bullock driven or with small power. But it will have to be ensured that the traditional labour employed in these areas does not suffer as a result.

Because of the inefficient *post-harvesting and processing techniques* this important income-generating activity is shifting to the urban areas. There is a need to develop efficient mechanical threshing, hulling and grinding equipment which can be operated on the farm itself to give a maximum return to the growers. The important areas in this field could be wheat grinding, which is carried out at the moment by small power (atta-chakki) wheat grinders, rice dehusking, and oil extraction from vegetable oil seeds. Factories producing semi-processed foods (canneries, bakeries, papad making plants etc.) are not suited to rural areas and attempts to introduce them have usually run into difficulties.

INDUSTRIAL PRODUCTION AND PRODUCT SELECTION

The next most important priority after food production is *cloth* which is used by every human being. As far as industrial production is concerned, there is a wide variety of products, ranging from supersonic planes and locomotives to pins and needles. A judicious selection of products from the list of new and old consumer goods will have to be made which can be produced in rural areas. The main problem will be whether a large enough market can be found for these products (which are mainly intended for the 10% of the population living mostly in urban areas) if, they are produced, on a large scale, in the villages. For example, at the present time only 5% - 10% of the population eats white granulated sugar, while everybody else, if at all, uses "gur", as a sweetener. If white granulated sugar were refined on a very large scale would it be possible to sell it ? This is irrespective of the fact whether it is desirable for such a large population to eat granulated white sugar. Similarly, toothpaste sells well, but if produced on a large scale by the villagers could it find an easy market everywhere and if manufacture were permitted on such a large scale would it not be responsible for an increased incidence of dental caries. The rural alternative to toothpaste is a twig of neem or some other tree, but these twigs do not give a substantial return to those who sell them.

Thus the other important question which the experts have to decide is what products should be selected for village industries, bearing in mind that they should give a substantial return to the producers and should also promote healthy consuming habits based on natural needs and not on greed and artificially stimulated needs. This fact calls for a shift in life style away from that promoted by Western industrial civilisation. There cannot be any doubt about some of the items already mentioned, notably cloth which is the next most important priority after food. It is very difficult to fix the third priority and we have to choose from products such as house building materials (not an item of regular consumption) and footwear (not worn by the poor, although derived from an important raw material, hide and skin being available in abundance following the death or slaughter of animals). As far as the food industry is concerned, milk product based on animal husbandry, vegetable oil and cooking fats produced from oil seeds and poultry etc. and certain items which do not come strictly within the compass of industrial production can be useful products. No one can dispute that these products are vitally important and are greatly beneficial to the health of the general population. A range of other consumer products could be listed their importance varying according to local conditions and requirements, e.g. wool products, jute goods, bullock carts (based on the age-old skill of carpentry), agricultural implements (bsed on the blacksmits's traditional art) pottery, pans and cooking utensils (based on the potter's ancient skill). Paper, sugar and cement are important consumer items of the present time which are manufactured only on a large scale. Fertilizers are also produced on a large scale. Power generation in the wake of the impending energy cricis. Because no attempt has been made to improve the technology of rural industries, these industries are migrating slowly to the urban areas. This has resulted in a reduction of the percentage of rural artisans and industrialists from 18% in the year 1910 to 7% in the year 1970.

TECHNIQUE OF PRODUCTION

Once we have selected the products according to our needs, altruistic ideals and their marketability the next most important step is to determine the technique of production. It is quite obvious that the technique which is used for large-scale production in urban areas would not be appropriate or suitable for production in the rural areas. Even reducing the size of the plant i.e. miniaturisation does not always serve the required objective. Research and development studies must be undertaken to determine the production technology for each and every item to the most important criteria is that the product should be comparable as far as cost and quality are concerned with the mass-produced products of large industries. It is true that production on a smaller scale gives an advantage, that if produced for the local market it saves costs in transport and in overhead management, tax and excise duty etc. Large-scale industry has a definite advantage in its economy of scale, in research and development expertise, in more efficient management, in easy access to financial and other infrastructure etc. These respective advantages can cancel each other out if the quality of small-scale products is good and costs are kept down. The other advantages of appropriate technology, apart from helping backward rural areas are : -

I   smaller gestation period as most of these industries can be established within six months of planning, whereas large industries take 3 years to 4 years, and the capital remains tied up;

II   Equipment does not have to be imported and most of it can be fabricated in small workshops and machinery can be brought from within the country.

After the research and development work has been completed for each technology, it is necessary to set up *pilot projects* in actual field conditions to test the economic viability and technical feasibility of the project under operational conditions.

The next important criteria for promoting small-scale production in rural areas is to have an efficient *organisational pattern*, the main hallmark of major industries. To simplify matters, it is better to have one institution which can serve many industrial production units in one field. The function of this institution would be to arrange suitable materials for the industry at the cheapest rate and of the best quality, buying at a time when cost are low and storing the materials for the rest of the year. In addition, the institution could arrange bank finance and credit, store product stocks and set up an efficient sales and marketing organisation. The research and development department could keep on working on new techniques, process and products so that as soon as the product stops selling the industry can switch on to diversify the product.

ORGANISATIONAL PATTERN

There can be many approaches for evolving a technique in its entire package

for small-scale production. Firstly it would be developed in its entirety
with the organisational pattern builtin, which can be sold to prospective
entrepreneurs who could then set up these industries on their own. By their
very nature and built-in constraints they would have to be situated in the
rural areas to be near to the raw materials. They would also have to be a
near to consumer centres to save expenditure on transport. The most obvious
products are sugar, cement and paper. The second approach could be to set up
a service centre, either controlled by the government, a cooperative society
or any other institution, which is prepared to provide an organisational
pattern to feed and serve a number of small producing units situated round
the service centre. The producing units would be too small to be able to
arrange for their own raw materials. The products best suited to this approach
are whiteware pottery, cotton spinning, rural oil etc. The second pattern
would place much of the productive activity in the rural areas and would be
more close to the ideology of production by the masses. The earlier model
would be owned by rich rural entrepreneurs who would employ workers on wages.
Obviously, the workers would not consider themselves to be partners in production and would be open to exploitation. However, the location of these
industries in rural areas will open up vast possibilities of activity in other
fields. It has been a matter of experience that these rural areas where
small-scale sugar and cement factories have been set up have grown in comparatively prosperous areas. Thus, both models have got something to offer to
rural development.

LIST OF PRODUCTS*
Regarding the list of products for starting new industrial activities in the
rural areas, there could be common agreement on the following minimum cate-
gorywise distribution.

1 By scaling down large scale manufacture -

a) sugar   d) cotton spinning   g) chemical fertilizer manu-
b) cement   e) jute spinning & weaving   facture
c) paper   f) wool spinning   h) soap making
                                               i) match making

2 By upgrading traditional village technologies -

a) handloom weaving

*A.T.D.A. has documentation booklets of all important projects and re-
search projects now going on at the association.

b) black smithy & carpentry

c) extraction of vegetable oil edible and non-edible

d) red clay and white ware pottery

e) tanning & shoe making

f) cereal processing like dehusking, grinding of wheat, paddy, spices, pulses etc.

g) gur and jaggery manufacture

3. Miscellaneous - power generation, domestic & household, educational services.

a) village power pool including solar appliances, windmills, mini hydel & bio-gas

b) environmental sanitation and drainage

c) village transport including bullock carts

d) animal husbandry & milk products

e) poultry and egg products

f) social forestry & forest based industries

g) land reclamation

h) educational materials such as exercise-books, fountain pens, chalk crayons, blackboards etc.

i) plastic goods such as toys, combs, cases and other practical items

A number of other items could be added to these lists depending on local conditions. It is estimated that if suitable technologies were to be developed for manufacturing the above mentioned items, direct employment would be created for at least 15 million people in the villages. Once that much employment were generated the economy would be greatly stimulated and indirect employment would be provided for atleast 50 million people.

POWER & ENERGY

The next most important priority for the rural areas of India, in some respects connected with industrial production, is the provision of power and energy for irrigation, household and domestic purposes as well as for industrial production. It is a matter of common knowledge that the plan for electrical generation and distribution has covered almost all the urban areas of the country in the last 30 years. Before independence, there were many district towns which had no electricity. Now there is hardly a block headquarter

which is not served by electricity. However, in most of the rural areas with
an electricity supply, the percentage is probably between 30% to 40% of the
entire villages of India. The electricity available is used generally for
energising tubewells only and is rarely used for any other purpose, perhaps
due to lack of demand or supply. It is quite obvious that if there is an increase in demand from all the villages which have been reached by electricity
for domestic connections and for industrial production there will be an acute
shortage of electricity which the present generation capacity will be unable
to meet.

Coal and oil have been used to generate electricity. India is acutely short
of oil, which has to be used for more important needs such as the production
of naphtha for fertilizers, transport etc. Good quality coal is also becoming
decreasingly scarce. Therefore it may not be possible to provide electricity
to all the villages by the conventional method of thermal power stations.
Moreover, the installation of transmission lines, transformers and sub-sta-
tions presents such a huge management problem that it will be very difficult
to manage the distribution of electricity. Already theft of transmission wires
and power is rampant in the rural areas and it is very difficult to control.
Moreover, electricity supply to the villages will always be controlled from
a central point miles away from the villages and they will have to depend on
one human agency for water, food and all essentials if they have to depend on
the Electricity Boards. Gandhi ji thought that it would be terrible to lose
the independence of the villagers in this way. It would be preferable for each
village or group of villages to have its own power house which it would be
able to manage, operate, repair and maintain and thereby exercise complete
control. For this purpose, an appropriate technology based on income energy,
and using locally available, replaceable materials will have to be devised
and developed. It has already been mentioned that a developing country such
as India cannot afford to generate sufficient power for the rural areas, which
are dependent on rapidly depleting resources of fossil fuel. The sources of
income energy   freely endowed by nature and available locally in the villages of India are as follows : -

1   Solar - About 300 days in a year is available in parts

2   Wind - Available at low velocity everywhere in India but in about 30% of
            the area it is available at a high velocity

3   Mini   Available in abundance in hilly areas and in low head available
    Hydel -

throughout in India in rivers, streams, rivulets etc.

4 Bio-gas -   Based on animal dung available in abundance in every village,
                but at present 60% is used as fuel for cooking.

5 Wood   -   Available in every village. Trees can also be planted in large
                numbers and used in a rational manner in such a way that they
                can be used as fuel either directly or by low temperature car-
                bonisation.

6 Others   -   Like geo-thermal, tidal, sea etc., which could be exploited
                by sophisticated and modern technology but not so relevant to
                the villages.

There is a need to carry out a detailed survey of the strength of all these
income energy sources to find out how much energy in terms of K.W.hr. they
are capable of providing for the villages. Next a suitable technology for each
of these resources has to be developed so that the maximum utilisation may be
achieved.

SOLAR

We have to find out whether it is possible to use solar energy for cooking in
every household taking into account the technical and social constraints.
Otherwise, is it possible to design solar power generators which can generate
electricity that can in turn be used for cooking, refrigeration, running of
engines etc.

A great international effort is needed in the area of solar energy alone to
increase the efficiencies of the absorbing and collecting surfaces already in
operation in many places. Similarly, work on utilising the solar energy so
collected has also to be done.

WIND

As regards windmills, there is hardly any design of windmills which can operate at low velocity. There is an established need for such windmills as nearly
70% of Indian villages fall in the low velocity region. However in the five
months from February to June there is intermittently high velocity wind which
can be used for pumping water, grinding grain and other low energy uses. How-

ever one of the greatest drawbacks of both solar and wind power is that they are intermittent and require a high capital expenditure per K.W.hr. for any attempt to harness them although there are no operating expenses once they have been set up. All these points have to be considered when designing equipment and contrivances for harnessing solar and wind energy.

BIO-GAS

The next most important resource available which can be profitably used for power generation is animal dung, which is readily available in every village where the animal population is usually nearly half the human population. Standard designs of bio-gas plants for the generation of mathane gas : from animal dung are already available. However, the low generation efficiency, the large volume of the plant, the high cost of installation have restricted their wide adoption in Indian villages especially by the poor farmers who do not have sufficient capital.

MINI-HYDEL

Mini-hydel plants will be useful especially in hilly areas where the people are already using the water energy provided by the poorly designed water wheels used for grinding wheat and other cereals. If a little attention were paid by technologists and designers to this problem it would be possible both to increase the efficiency of the water wheels and to use them for electricity generation and taking electricity on wires to remote villages which are not served by streams of waterfalls.

WOOD

Wood is the primary kitchen fuel in village households. The villagers pick up the supply of their fuel wood either everyday or periodically and store it for use in their kitchens. However, no planned effort has been made to plant trees on a large scale households. The fuel like the fossil fuel crisis has now assumed serious proportions. This is due to lack of elementary foresight because trees take only a few years to grow, not millions of years like fossil fuels. Even now if a certain percentage of the land area in each village were reserved for tree plantation, this crisis could be averted and a regular supply of fuel wood could be assured. The most efficient method of burning wood has also to be determined by the technologist and scientist. At present, the burning of fuel and cooking is carried out in a most efficient manner resulting in loss of valuable calories at the time of the transfer of heat from the burning fuel to the cooking utensils can solve this problem and reduce the loss of heat to a considerable extent.

Wood can also be used in a more efficient way by the process of low temperature carbonisation. This would provide a supply of cooking gas and charcoal for fuel. All these methods have to be investigated to ascertain how much contribution each, namely solar, wind, hydel where available, bio-gas and wood can make to a village power pool.

PHYSICAL AMENITIES IN THE RURAL HOME

The most important activities for the rural areas have been covered in the earlier paragraphs which dealt with the production of food and the setting up of industrial production activities for income generation for the rural areas and also for surplus formation for other investment and the supply of energy for these purposes. Now we have to consider other areas for improving the quality of life in the rural areas and providing physical amenities and services which are lamentably lacking in most of the villages. These can be enumerated as follows :

1	Rural Water Supply	– Most essential for healthy living, elimination of worms, germs causing dysentery and other diseases
2	Sanitation	– This would involve setting up of sanitary latrines in the household and drainage of water etc.
3	House building	– To improve the quality of shelters in the rural areas and to develop suitable materials for this purpose which are more durable, water proof and fire resistant.
4	Construction of roads	
5	Health care	
6	Household & domestics	– For cooking, washing of clothes and utensils,

supply of hot water etc. This has been covered to some extent in the earlier chapters.

| 7 | Education & culture | – Education is one of the most important requirements to change the attitude of the people and to give them knowledge about the best methods for improving their environment and to make best use of the material available around them. The pattern of education developed by the industrialised countries was such that it has given false hopes and expectations to people of the affluent and extravagant life being generally led in the western countries. This has alienated the educated rural folk from their roots and they are not able to put in their best for the development of the local environment. A proper system of education has to be developed by the rural people themselves which can help them to carry out the activities as mentioned in the earlier chapter to the best of their capabilities and thereby to ear an sufficient income for themselves and also to improve the quality of their life and environment. |

These are the only methods through which the 66% of the people of the world living in the rural areas mostly of the developing countries in 2 million villages can be salvaged, otherwise there is no hope.

INDUSTRIALISATION IN WESTERN COUNTRIES

'Science' is more or less neutral, but technology, being application of science for the practical use, is not neutral and has a strong class and social bias. The technology of war time is too well-known, but even the technology of peace time has brought enough disaster for the mankind. Unfortunately, science, which started as a branch of natural philosophy, has been commercialised in recent times. There is nothing like pure knowledge. What can be proved theoretically, must be done practically. If the scientists have shown that it is possible to split atom with release of tremendous energy, it becomes necessary for the technologists to manufacture nuclear bombs and nuclear power stations. If the scientists show that it is possible to counteract the forces of gravity, the technologists immediately get together to enact the greatest circus in the universe by sending man on the moon and arranging acrobatics in space. (Perhaps, we could leave some of the discoveries of science in the realms of pure knowledge and just for teaching purposes). As a result of this the science is becoming more and more dependent on technology in modern times.

This alliance between science and technology for the last two centuries has brought a level of affluence in the western industrialized nations, which is beyond imagination and comprehension. We all know too well that at one point, it brought untold misery and unhappiness in its wake. The advent of industrial revolution led to great exploitation of the working class in the initial stages. This state of affairs continued for generations before wealth in the industrialised countries started percolating down towards the lowest social strata and the poor working class have been able to get necessities of life and some of the luxuries. Yet, by no means does every one in the industrialised world live in comfort. No efforts have been made to consolidate the gain of technology for everyone. The innovators are more concerned with achieving supersonic flights than improving living conditions. With the invention of radiant and central heating, the technology of heating interiors has reached near-perfection; but still there are thousands in the industrialised nations, who are unable to pay their fuel bills in winter and die from cold. The fruits of high technology are available only to those, who can pay for them.

The search for new methods of production to achieve affluence has also started a mad scramble for the world's resources in material and energy. More often than not, we hear cries of depletion of resources and energy. Man's unappeasable appetite for goods and profits have exhausted vast deposits of world's resources in no time.

A second crisis gripping the world today, due to the indiscriminate use of technology, is that of pollution - the danger to ecology, the balance of living world around us. As long as the pollution and other ecological disturbances, wrought by industrial production were small, nature could absorb and assimilate them. Now even the vast ocean, the total biosphere and the whole of earth have become unequal to absorb the huge amount of garbage, waste, effluents, fumes and radiation released into them.

71

The modern technology has created serious crises, namely the crisis of depletion of resources and energy, the crisis of pollution or destruction of environment and ecology and disturbances of the balances of nature, and also
the destruction of human values and alienation of the human being from the
process of production. Was it a sensible thing to have based the entire economy of the western developed world on one single thing, which they do not
produce themselves, i.e., oil. Even if the oilproducing countries do not stop
oil export to industrialized nations, there are all indications that there
will not be a drop of oil left in world reserves by the turn of the century.
The second important crisis, which is afflicting the industrial nation is the
pollution of our environment, by the production of garbage, waste, effluents,
liberation of fumes and gases in the atmosphere, radiation etc. It seems that
modern civilization will be buried under its own garbage heap. This situation
has arisen out of the grand success of the industrial society, as if the ants
have discovered their own insecticide!

Industrial society has used the human substance as its raw material too. It is
for this reason that there is a crisis of youth in the rich nations. Some of
the best of those youth say, 'A course on all your houses, we don't need
them' and unilaterally discard the vulgar use of the products coming from modern industry, and all sort of organisations like the New Alchemy Institute
'Friends of the Earth', 'Movement for a New Society' are coming up. They are
rebelling against the established order, the very basis of the industrialised
society.

It is profoundly dismaying that one of the major efforts of the developed
countries in most world War II years have been to destroy the concept of
enoughness, which existed in many religions and philosophies of the non-wes-
tern world. Development theorists and businessmen saw correctly that the economic growth could not be attained without the achievement of new desires for
goods and services. We are only now perceiving that an insatiable desire for
additional goods and services is infeasible in ecological terms and all too
often destructive in human terms as well.

We shall have to learn what it means to live in a world of enoughness. We
must break our present behaviour, which is based on the belief that more is
necessarily better, that consumption provides satisfaction in itself. Ghandi
said more than 40 years ago that "There will always be enough for everyone's

need, but not enough for one man's greed".

This situation has let to vigorous protest in the developed countries themselves. Ivan Illich in his series of book, 'de-schooling society', Energy and
Equity', 'Tools for Conviviality', 'Medical Nemesis' has decried the present
institutions and calls for 'counter-research' which can be carried out only
in a 'de-schooled' society. Peter Harper, writing on 'Notes on Soft Technology for UNESCO has spelled out the ingredients of sort technology, based on
which a 'counter-culture' has to be developed. The tallest of them all, late
Dr. E.F. Schumacher, in his brilliant books, 'Small is Beautiful' and 'A Guide
for the Perplexed', has given a call for adoption of a 'technology with a
human face' by both developed and developing countries, the lead for which
could be provided by the Third World, which has not been so debased by the
concept of greed and profit and has still time to take to an absolutely new
path for development and progress. Writing on the contradictions of the modern
civilisation, he wonders, that if an ancestor of an ancient time came to the
world, what would he be more astonished at : the skill of our dentist or rot-
tenness of our teeth, the progress of our medicine or overcrowding of our
hospitals, the speed of our transport or the length of time and discomfort of
our journey to and from our work, our ability to land man on the moon or our
inability to find work for the workless, the efficiency of our machines or
the inefficiency of our system as a whole. The most unfortunate thing is that
our predicament is not due to our failures, but due to our successes.

Robert Theobald in his book 'Failure of Success' says that unless the world
moves into new era, which he calls 'Communication era', there is no chance
for our survival. The developed nation has to move from the industrial era
and the developed countries can move directly from agricultural era, without
going through the industrial era. He outlines that "the keynote of the culture
in communication era will be an understanding of process and acceptance of
cooperation in their relations with nature and the environment. Those living
in industrial era have tried unsuccessfully to replace these values with force and competition. He says 'The communication era will be a world of diversity, if we succeed in bringing it into existence. Our sterile monoculture
of kids and crops must necessarily give way to synergetic interactions of all
living things. We now know that environments are dynamically stable, when
they contain large number of organism style and culture. The drive towards
equality and similarity was necessary to the functioning of the industrial

era. We must break down the uniformity and create a world of profoundly
unique individuals grouped together in widely diverse community.

It may be appropriate at this point to quote from Mahatma Gandhi, the Indian
sage and visionary, who could see far beyond others and had sounded warning
signal against industrial society, more than 40 years ago. He wanted India to
develop a new technology to make villages 'little self-sufficient republics'.
He knew that the method of industrialisation and development would generally
by-pass the villages, where 80% of the people lived. "If villages perish,
India perishes too", he declared, "but whether for good or bad, why must India
become Industrial in the Western sense? The western civilisation is urban ...
A big country with a teeming population, with an ancient rural tradition,
which has hitherto answered its purpose, need not, must not copy the western
model". He repeatedly warned that industrialism is based on exploitation.
Writing in Young India, in 1931 he said "Industrialism is, I am afraid, going
to be a course for mankind. Exploitation of one nation by another cannot go
on for all time. Industrialism depends entirely on your capacity to exploit,
on foreign markets being open to you ... a vast country like India, when it
begins to exploit other nations ... And why should I think of industrialising
India to exploit other nations".

He was not against machines, as he said on many occasions. In the year 1924,
he said "I have no design upon machinery as such; I would prize every invention made for the benefit of all, welcome the machine that lightens the bur-
den of millions of men". He made the spinning wheel a symbol of such a machine.
He was very much against labour-saving and automatic machines and said, "Men
go on saving labour,till thousands are without work and thrown on open street
to die of starvation. I want to save time and labour not for a fraction of
mankind, but all; I want concentration of wealth, not in hands of a few, but
in the hands of all. Today, machinery merely helps a few to ride on the back
of millions. The impetus behind it all is not the philanthropy to save labour
but greed. It is this constitution of things that I am fighting with all my
might." In conclusion, I may ask this question : what should be the ingredients of the new technology, which should lead to a life of peace and permanence for the whole world and bring a better harmony with nature. It should
be small, so that small persons should also be able to make their modest contribution. It has become too big and gigantic. It has now become too complicated. It should be decentralised and widely dispersed; it has become too

concentrated and centralised. It should be less energy and resource-intensive;
it should be pollution-free, exploitation free and non-violent, in harmony
with nature, rather than bludgeoning it all the time. This modern technology
has got no long-time future. It cannot encompass the whole world. The time
has come to ask the question : which is better, a Rolls Royce with an empty
tank or a bicycle? If it is at all possible to feed the whole mankind, then it
will only be possible by ecologically and biologically sound methods, through
the careful recycling of all organic materials. This will be a 'technology
with a human face' in Schumacher's phrase. This will be a technology of self-
reliance for the developing countries and a technology of limitation for the
developed countries. I hope that research in science and technology would
keep to these requirements as far as possible for the benefit of mankind.

STATEMENTS

- Appropriate technology should be "a technology with a human face". It must be a technology of self reliance for the developing countries and a technology of limitation for the developed countries.

- On the subsistance level traditional technology is a very important issue much of the skills there seems to vanish. This must be a major care of organisations working in the rural areas.

- Development of a technology for the rural areas must be development of a technology suitable to create jobs in the rural areas.

- In the development of appropriate technology hardware it is important that designs can be duplicated and used else where.

APPROPRIATE TECHNOLOGY IN SRI LANKA

*Calyanatissa A. Gunawardhana**

INTRODUCTION

The period between 1970 and 1977 saw the introduction of a large variety of
economic experiments in Sri Lanka. Among them, the most controversial was the
Divisional Development Councils (DDC) programme launched by the Ministry of
Planning & Economic Affairs, where, for the first time an attempt was made to
accomplish regional development through popular participation utilizing appropriate technology.

Here, popular participation meant the joint effort of the Government officials
and people's representatives and appropriate meant, for the most part, indigenous technology. The programme encompassed a wide spectrum of disciplines
and was able to generate employment on an unprecedented scale. The viability,
however, depended on a number of controls, primarily, the restriction on imports.

The programme found favour in the less developed areas of the island where
the participants were totally committed towards the project as against in urban areas, where it failed to harmonize with the more sophisticated environment.

Due to its partial success and its viability resting on stringent controls, it
did not find unanimous support in the planning circles. Its adherents worked
with an almost religious fervour whilst the programme was in force but had
to face realities of free enterprise which led to its abandonment in 1977.

This study outlines the events that led to the identification of technology

*A.T.G., Sri Lanka
the paper is based on observations on the Regional Devlopment Programme of
the Ministry of Planning and Economic Affairs 1970 - 1977

for the programme and is not intended to be conclusive; it is a part of a
large subject now termed as appropriate technology. It is hoped that the observations to follow would definitely contribute towards the derivation of
better methods of accomodating concepts of appropriate technology in development programs, particularly in the context of an impending energy crisis.

HISTORICAL BACKGROUND

It is now clearly established that evolution of technology is directly related
to the necessity, environment and availability of resources. Within this process there appears to have been a constant transfer, or, at times, disruption
of technology, either by migration or following occupation by conquest.

A brief comparison between Sri Lanka in relation to countries in Europe within
recent history confirms the validity of this statement when around the period
between the 15th and 18th century, technology was either at par or even in
some instances superior to those in Europe. For instance, in hydrology - water
conservation and its management; in ceramics - manufacture of hydraulic ce-
ments and glazes; in metallurgy - extraction of iron, copper, gold and silver
and manufacture of articles thereof, are some examples where technological
advancement was well known. The writings of European chroniclers confirm this
view.

*"... The advance from iron to steel was rapid and Sinhala weapons partially
carburised into steel have been excavated from sites 2.000 years old (Hadfield
1911), while the Vaijayantha written prior to the arrival of the Europeans in
the East, gives some idea of the weapon making that evokes their admiration.*

*Pyrard (1679), a Frenchman, praises their skill in making and ornamenting
arms which he says were esteemed the finest in India; adding 'I never thought
they could show excellence in fashioning arquebuses and other arms; more beautiful indeed are those in workmanship and ornament than those made in France.'*

*Liuschoten (1592), a Dutchman, writes that they make the fairest gun barrels
to be found in any place, and that these shine like silver. A Dutch plan of
the Royal Palace at Kandy in 1765 showing the armories for different types of
weapons and quarters for the armorers proves the great attention paid to
weapons by Sinhala Royalty ... "* [1]

A representative example of the Sinhala metal workers art of a bygone era
may still be seen at the Rijks Museum in Amsterdam - an extremely ornate bell
mouthed canon which once belonged to a General of the Sinhala army (circa
1745).

*"...the people of Ceylon work with considerable skill and taste in gold and
silver, and with means that appear inadequate execute articles of gold and
silver jewellery that would certainly be admired, but not very easily imita-
ted in this country (Britain). The best artists require only the following
apparatus and tools : a low earthen pot full of chaff or saw dust, in which
he makes a little charcoal fire; a short earthen tube or nozzle the extre-
mity of which is placed at the bottom of the fire and through which the artist
directs the blast of the blow pipe; two or three crucibles made of the fine
clay of the ant hills, a pair of tongs, and anvil, two or three hammers, and
a file; and to conclude the list, a few small bars of iron or brass about two
inches long, and differently pointed for different kinds of work. It is as-
tonishing what an intense little fire more than sufficient strong to melt gold
and silver can be kindled in a few minutes. Such a simple forge deserves to
be better known; it is perhaps, even deserving the attention of the scientific experimenter, and may be useful to him when he wishes to excite a small
fire larger than can be produced by the common blow pipe, and he has not a
forge at his command .... the blacksmiths of Ceylon are not behind their
brethren, the jewellers, in the simplicity of their apparatus ...."*[2]

However, towards the 18th century, in Sri Lanka, as in most countries in Asia,
the pattern of technological advancement, within the parameters already described gave way, due to a number of factors which are extremely complex. By
the turn of this century not only there existed a technological disparity between Sri Lanka and Europe, but some of the technologies which had been common knowledge were forgotten.

POST WAR DEVELOPMENT

In the 1950's Sri Lanka boasted of as having one of the highest standards of
living in Asia, second only to Japan. There was a high degree of specialisa-
tion on three plantation crops, namely, tea, rubber and coconut and probably
due to the complacency which arose out of a temporary prosperity no significant attempts were made either at diversification of the agricultural sector

or at industrialisation.

Over the years to meet the demands of an increasing population, there was a
corresponding increase in the imports of food and consumer goods without a
corresponding increase in production, resulting in adverse balance of payments
problems which led to a curb on imports. The increase in population also disrupted the agrarian structure and resulted in the subdivision of holdings
which led to landlessness and unemployment.

The early development plans were aimed primarily at import substitution, and
one witnessed a weird concept of industrialisation, where industries were
set up to "Produce" items the imports of which were restricted. For instance
a large number of units were established to manufacture confectionery which
depended on imported sugar. Similarly the units which manufactured radios,
imported sets in knocked down condition (to 3 or 4 components) which were
assembled locally. The manufacture of razor blades meant the packing of the
finished product which was imported. Ironically, there were restrictions in
the establishment of such industrial units and it was no secret that corrupt
practices prevailed in the registration and allocation of foreign exchange
to a privileged few to import the 'raw materials'.

In paralell, the Government established industrial corporations in areas
where the private sector showed no interest at the time, 'in the interests
of greater efficiency and higher production as well as in the interests of
a society as a whole'. The comments of a Minister of Finance on State owned
corporations deserves attention.

" ... What has been the experience? One cannot but admit that this form of
organisation has not come up to expectations. As so often happens, people
chosen to the boards of directors have not always been men of the highest
calibre or men whose integrity is beyond question. Too often, they have not
been men with any particular capacity to understand the problems connected
with that venture over which they reigned. Unlike boards of directors of
companies in the private sector they have no financial stake in the corporations in the sense that directors are shareholders of companies. They
did in a sense have a stake in the finances of the corporation. They were
interested not in the share they put in but in the share they pulled out.
Boards of Corporations very often degenerated into instruments for appoin-

ting people. Members back-scratched in order to ensure that their nominees got into these corporations. This was foreign to the original intention. These boards have now become the mudliyarships* of our time. They are the means of recognition of political service. Necessarily, therefore, the executive and the administrative staff are concerned less with the efficient performance of the enterprise under their charge than with placating the whims and fancies of the directors of the board. It has gone by default because directors could do no wrong. One is not surprised therefore, that some of these state-owned corporations have become a liability rather than an asset ..." [3]

The failure of state owned industrial corporations in an accepted phenomenon in the free world and recent reports on countries where extreme forms of planning & management exists the situation appears to be simular.

The Government was virtually drowned within its own adminstrative machinery to a point which led to economic stagnation and decay, which was further aggravated by the restrictions imposed on the private sector. There were numerous industrial corporations, departments and statutory boards and there was little or no coordination among the state agencies.

For example, there were over 20 state agencies under six different ministries directly connected with the rubber industry and to have formulated a development programme that could have been implemented was indeed an exceedingly difficult task.

A significant feature of the 1965 - 1970 period was the creation of the Land Army by the Government, where unemployed youth were mobilised to execute public works, as done in China. This attempt at solving the unemployment problem was unsuccessful and the regimented workforce was disbanded in 1970.

It was in this context, that with a change of Government in 1970, that the DDC programme was launched at the beginning of 1971 to

---

*A title confered by the British during colonial occupation. No powers were attached to the title. The recepients were given a helmet, a black coat with number of brass buttons, a sash, and a small sword with an ornate handle which they wore on ceremonial occasions. The conferment of titles and honours were abolished in 1956.

" ... bring together peoples representatives and Government officials for the
common task of regional development. ... The councils are expected to examine the land and natural resources and raw materials available locally and
to devise projects which will provide employment opportunities for the people..
.. in other words, the Development Councils are given an opportunity of producing a miniature plan for their areas ... " [4]

About the same time the Five Year plan was drawn and it was meant to serve as
a guide for the 5 year term of office of the Government which followed (1972 -
1976). The philosophy of the DDC programme was expressed in no uncertain
terms. The plan issued in November 1971 stated that

" ... the DDC's will be the main link between the network of Government Agencies on the one hand and the local community and its representative institutions on the other. The Councils consist of Government officials and of representatives of Institutions of co-operative societies, cultivation committees, peoples committees and village committees. The function of these
Councils includes the formulation of Development projects and the preparation
of development programmes for their areas. They will also assist in the coordination of development activity and the review of plan implementation in the
Development Council areas ... " [5]

The State agency that was expected to provide technical consultancy services
was the Industrial Development Board, an agency which came under the Ministry
of Industries and Scientific Affairs.

" ... The IDB was created by the International Labour Organisation as a part
of its programme for boosting fresh avenues of employment. Originally it was
set up on the lines of the Village & Khadi Commission of India. Its first
Chairman, was a member of the all powerful civil service*. Records at the
IDB will prove beyond doubt that he revelled in showmanship fostering com-
pendiums and reports costing over 8 million rupees of tax payers' money ...
(which) turned out to be a good only on paper, for they were later found to
be not at all practical; often massive slices of the reports were copies of
Indian deliberations ... "

*A tradition set up by the Britisch during colonial occupation. Senior
 Govt. officials, like Revenue Collectors fell into this category.
 Civil Service was abandoned in the early 1960's.

" ... Members of the public however, are compelled to go to this body to get approval for their own industries brought out on their own initiative, only for the purpose of obtaining their approval and registration ... "

" ... work on the IDB showed that they had got an open impetus to corruption and although such corruption could have been checked, instances can be found where policy makers virtually aided and abetted corruption in varying degrees.. ... in certain cases of todays industries the IDB helped the continuance and preservation of monopolies in favour of certain industrialists. Glaring instances can be brought about to justify this statement ... "

" ... The Industrial Development Board has not justified its existence so far. It has no right to carry on any further. What industries the country can boast of today have envolved solely on the enterprise of private sector initiatives ... "[6]

These comments indicate the inadequacy of a specialised agency set up by the ILO and the relevance of such institutions in industrial development.

These factors led the planners to look towards traditional (or) indigenous technology; they were not quided by nationalism or any other political thought, as often claimed by some critics on the DDC programme.

IDENTIFICATION OF TECHNOLOGY

The choice of technology as seen from what has been already described, was rather of circumstance than that of deliberate planning. Neither the large industries nor the Government of the day accorded the scientific and technological development the priority it demanded an right along failed to appoint scientists and technologists to positions of executive authority. Parliamentarians and Government officials had vague notions of science and technology. High priority was given to education in pure science, which bore no immediate relevance to the developmental strategy of the country. The concept of technological education was misunderstood and Technical Colleges received low priority.

As a consequence of this lopsided educational policy there were a large number of science graduates who could do nothing more than teach the subjects they learnt at the University to school children. Worse still, the curriculum

offered by the Technical Colleges served no more than to produce technicians
to service institutions inherited from the colonial Government.

Of the category that received specialised education in the developed countries, only a handful were able to fit into areas of their speciality on
their return. A large majority engaged themselves in administrative work
which very often bore no relevance to their field of study. Many had their
training in sophisticated factories in Europe and on their return they found
that factories at home did not match the techniques of production, which were
often out of date; this led to frustration and compelled them to find employment in developed countries.

The research institutes conducted research in isolation and had little or no
contact with industry. When efforts were made to provide extension services
to existing industry the results were marginal.

*"... The type of product and the volume of production, in general, demands
only a fundamental level of technology and in majority of the units the services of specialised technologists have rarely been found to be necessary.
The foundations of present industries were laid down by untrained people, some
of whom, to their credit have developed ingenious techniques of overcoming
inadequacy of machinery and equipment they possess. It is primarily due to
this reason that they have developed a certain built in resistance to coun-
selling and advice (probably for good reasons). To overcome this a new dia-
logue is necessary, and will have to take into consideration even the cultural traditions of the people. For instance, in one factory, where there happened to be a breakdown of a mill due to the fatigue failure of a gear wheel,
one of the possible reasons given by the industrialist was that 'the mill
had not been installed on an auspicious day' ... "* [7]

*" ... Another good example of fruitful interaction between intellectual and
manual workers comes from Sri Lanka, where metal working and tool making are
ancient arts. University students of metallurgy were sent to share their
knowledge with the blacksmits. To their total surprise, the students discovered that the Bethlehem Steel colour-for-temperature charts they brought
along corresponded exactly to techniques local smithies had always used. Now
that the initial shock has passed the students are learning at least as much
as they teach ... "* [8]

In this context, the avenues open were indeed limited and there was no alternative to indigenous technology in the development programme. The areas of
activity were numerous and Light Engineering, Geological Resources (ceramics & phosphate fertilizers) and Ayurveda (indigenous medicine) were given
high priority as large numbers were involved in the three areas described.
This study confines itself to discussing some aspects of the development activity of these within the DDC programme.

LIGHT ENGINEERING

Among the artisans in Sri Lanka, the blacksmiths represent a highly skilled
workforce, and their traditions in the art of turning out iron and steel implements dates back to centuries. Today, they compose the nucleus of the
light engineering sector, and play a significant role in the national economy.
The DDC programme took cognizance of this fact, as it was one source from
which one had to draw skilled personnel for specialised activities. It was
therefore vital that this group was not only sustained with state patronage,
but high priority was attached to development activity in this sector.

The setting up of an organisation to develop this sector by a state agency
was first conceived in late 1972 at the Ministry of Planning & Employment, as
it was found that none of the existing agencies at the time were equipped to
participate in its development. In Juni 1974, the Ministry of Planning & Economic Affairs gave a further impetus by setting up an apex organisation with
the fundamental objective of

- setting up light engineering societies at district level
- procuring raw materials
- assist marketing of finished product &
- sponsor training programmes.

The apex organisation, the Light Engineering Union was the first development
agency within the co-operative concept. It was to be self sustaining as
against the IDB or the Department of Rural Development which depended on out-
right grants from the State.

The official status of the Union was quite clear at the time it was set up.
It was a DDC project and the Ministry of Planning & Economic Affaires directly participated in its activities. Since its inception, Ministry officials

directly participated in the management as a part of their official duty; the Director, Regional Development represented the Secretary, as its Chairman; a Deputy Director served as the General Manager of the Union. Most of the executives were officials attached to the Ministry.

In its first stage of activity, the village blacksmiths, numbering around 17.000, who were dispersed in the various districts were organised on the basis of a co-operative. 108 such co-ops were organised. These co-ops manufactured tools and equipment utilising relatively simple techniques.

In its 2nd stage of activity, about 19 medium scale co-ops were established in close proximity to urban areas. These employed comparatively modern techniques and produced tools and equipment of a more elaborate design. These co-ops played a significant role in the diffusion of technology in the respective areas.

In its 3rd stage, Technical Service Centres were to have been set up in the various districts and they were to have played a complementary role in the development of this sector. This stage was not accomplished, though the Government of India gifted machinery and equipment for 6 such units.

The supply of raw material was mainly from the Railway Dep., Transport Board workshops, Port Commission workshops and the Steel Corporation, where large quantities of iron, steel, copper and brass were lying unutilized.

The marketing programme was directed towards the public sector where the Union enjoyed a privilege granted by the Government. By a Treasury circular, all Departments and Corporations were directed to purchase products from the co-ops without resorting to competitive bargaining.

The largest production workshop in Sri Lanka, was purchased by the Union in 1974, at a time when the owners were compelled to close the factory and retrench the workforce due to heavy operational losses. The project was financed by the Peoples Bank and was jointly managed by the Union and the Ministry. This served as the Technical Centre and a number of developments in Light Engineering and alternate energy were conducted at this Centre.

It is relevant to record that this Union pioneered the development of the 2 wheel tractor and a suitable internal combustion engine to power the same.

" ... The internal combustion engine, the principle of which were laid down
by Beau de Rochas (1862) was developed by Otto, fourteen years later in
1876. This engine has remained very much the same to this day. Under the 2-
wheel tractor development programme of the Ministry of Planning & Economic
Affairs a frantic effort was made by the more dynamic officials of the Regional Development Division to obtain the assistance of 'engineers and scientists' to develop a small engine to power a 2-wheel tractor ...

... Then came the excuses. In desperation the enquiries were directed to lesser mortals and Wickremasinghe (true to his name) in a small workshop, (where
the only lathe was one built by himself) at Minuwangoda turned out the first
internal combustion engine in this country. When my colleague, D.L.O. Mendis,
who was President of the Engineering Section of the Association for the Advancement of Science proudly presented Wickremasinghe and the engine at a
meeting the members present were taken by surprise ...

... Padmasiri Dias, came out with another a few weeks later, where six months
earlier in the same workshop, Hector Wijetunga smelted steel from Seruwila
iron ore. Sanath Ranatunga has since perfected a two wheel tractor to suit
local conditions : already 10 of these have been tested and large scale production will soon commence in a workshop being built under the DDC program." [9]

The work done by its Research & Development Unit was indeed creditable, particularly in alternate energy development.

" ... for instance, in the event of failure of fuel supply (petroleum) this
group came with plausible alternatives to meet an emergency. The use of
Muthurajawela peat to generate electricity was worked out; experiments were
conducted to use it for the recycling of metals. The installation of small
hydrel stations were mapped out; windmills were developed (Savonius & propel-
ler types); and electric car was developed; a rice processing machine which
utilised rice hull as a fuel was perfected; bio-mass gas generators to run
vehicles were developed ... majority of them worked outside their area of
speciality ... " [10]

Though the policy at Planning level was to foster and develop light engineering sector, and thereby provide secure employment to a large group, there
appeared to be a lack of coordination among the Government agencies. For instance, the Ministry of Industries through the State Hardware Corporation to

a significant degree undermined this effort.

For instance, the State Hardware Corporation, despite having facilities to produce sophisticated items, which could not have been turned out by village blacksmiths, continued to produce the very same items that the blacksmiths were turning out and sell at competitive rates. The result of this was that the Hardware Corporation suffered colossal losses and this exercise did neither benefit the Corporation nor the light engineering sector.

The development work connected with the 2-wheel tractor and the alternate energy programme came to an abrupt end in 1977, when the Union terminated its activities.

CERAMICS

The traditions of ceramic workers in Sri Lanka, in the manufacture of earthenware pots, dishes and other clay products (bricks & tiles), as in many other lands, dates back to centuries. No distinction can be drawn between ceramic workers in Sri Lanka and in any other country, except for the fact that Lankans managed to survive against all odds and preserved their art without patronage from the state.

In the 1960's a giant corporation set up by the State to manufacture ceramicware using highly sophisticated automatic machinery enroached into their area of activity and cornered the traditional ceramic workers to much hardship.

An aluminium utensil factory launched about the same time mass produced pots and pans and disrupted the traditional demand for local earthen ware.

Maha Yala, in Bulanthsinhala, in the Kalu Ganga (river) basin is a village where about 500 families were engaged in the manufacture of earthenware. In 1973, they were eking out an existence and were dependent on the free rice ration given by the Government. The production was scanty and the demand was marginal, and very often ruthlessly exploited by middlemen. They procured the raw material (clay) by purchasing a few square feet of rice field (lying fallow), at a time and winning the clay using hoes. The payment being for the surface area, the workers, tunneled their way to win more clay. There had been many instances where people had died in the process. Many had been compelled to give up their trade, joining the ranks of the unemployed.

The officials of the Ministry of Planning & Economic Affairs who visited the village discussed the problem with the representatives of the artisans and as a first step an abondoned rice field was purchased to provide the raw material - clay, on a credit basis. A centre was set up, where a pug mill and a firing kiln was installed. The de-aired and screened clay was distributed from the centre. The sun dried earthenware were brought back to the centre for final firing. The marketing arrangements were made through the co-ops in the district.

Restrictions were imposed on the aluminium utensil manufacturing unit, an automatic plant which employed around 20 workers.

Within a period of 6 months or so, the whole village was active and on the average the ceramic workers in the village were better off than their counterparts at the Ceramic Corporation.

Similar projects were set up in various parts of the island and the ceramic workers were organised into co-operative societies, and as to the light engineering sector the concepts of management were similar.

The assistance of the Ceramic Corporation was sought to upgrade the technology, but the response was poor. While this programme was on, the same Corporation went into collaboration with a large multinational firm and set up an even a larger unit, 'to earn valuable foreign exchange'. The investment to generate a unit of employment at this project was 160 times more, as against the traditional projects.

AYURVEDA

" ... In the sphere of health services, the share of public expenditure amounted to 2% of the Gross National Product and about 7% of the total expenditure. The development of health services has always meant an increased expenditure on imported drugs and equipment and very little or no emphasis had been placed on research, development and manufacture of pharmaceuticals from locally available materials ... " [11]

The increase in expenditure over the years was not only due to the increase in population but also due to the gradual displacement of the traditional systems of medicine by the western system, which depended on imported pharmaceuticals.

The programme to rehabilitate Ayurveda was launched at the end of the Five Year Plan period, and significant steps were taken to consolidate and strengthen the system. The ultimate goal was to integrate both traditional ayurvedic system and the western system as done in a number of countries in Asia.

One area where efforts materialised was in the introduction of acupuncture in Sri Lanka, following a decision made by the members of the panel of indigenous medicine of the Ministry of Planning & Economic Affairs in February 1974.

The practice of acupuncture was non-existent though there were numerous palm-leaf manuscripts, some of which were centuries old. The Government of the Peoples Republic of China assisted in this programma, and three physicians were trained in China, who on their return were able to train a large number of both Ayurveda and western physicians.

The Ayurvedic Research Institute initiated a research programme into acupuncture and comparisons between the Chinese and Sri Lanka system were made.

In the programme for the rehabilitation of Ayurveda the co-operation of the Netherlands was sought and as a result the Delft University of Technology and the Utrecht State University through the Appropriate Technology Center offered assistance to upgrade process technology in Ayurveda, which had been untouched for centuries.

In the absence of the DDC programme, this project is to function as an independent project under the aegis of the National Science Council of Sri Lanka. The Ministry of Plan Implementation has attached high priority to this project as the drugs in Ayurveda are based on plants and herbs, which could be grown in Sri Lanka and being a renewable resource, it would have far reaching benefits.

EVALUATION

The evaluation of the DDC programme, in terms of economic achievements is an extremely difficult task, as a major part of the activity was rehabilitation of artisans. The capital invested on the programme as a whole, and the return on investment, the employment generated and the investment to generate a unit of employment has been worked out and it has definitely surpassed the achievements of the state corporations.

The various projects were commissioned under different conditions over a 3 year period and as a number of major reforms took place about the same time, e.g. Land Reform, the importance and significance of the DDC programme, was obscured and at times, it was suspected to be politically motivated. This was probably one reason which led to its abandonment in 1977.

The DDC programme succeeded as an instrument for the revival of traditional technology in Sri Lanka. It also afforded a large number of individuals from various backgrounds to work outside their area of speciality, in development activity on a common platform.

TECHNOLOGY

The identification of technology for the DDC programme, as stated earlier, was one of circumstance than that of deliberate planning. To slide down the scale of technology is prompted either (1) shortage of skilled manpower or (2) problems associated with commercial viability (3) problems associated with the supply of raw material or by far the most important (4) the difficulties in obtaining cheap and abundant energy.

Some of the technologies, which were abandoned nearly a century ago, viz.air-motive engines (Stirling 1827, Crossley 1828), Belier hydraulique (1800), electric passenger vehicles (1890), have made a comeback. Even the horse carriage, presently used by the royalty for ceremonial occasions, may once again become a popular vehicle.

Technology abandoned in the past is becoming the technology of the future.

REFERENCES

1 *Deraniyagala, P.E.P.; Sinhala Weapons & Armor,* Journal of the Royal Asiatic Society Vol. XXXV.

2 *Engineers Encyclopaedia, Vol. I, 1835.*

3   Perera, N.M.; *Address to the Ceylon Association of Advancement of Science, Dec. 1971, (Minister of Finance 1970 - 1975).*

4   Gunasekera, H.A. De S.; *Development through DDC projects 1971.*

5   *Five Year Plan (1972 - 1976), released by the Ministry of Planning & Economic Affairs.*

6   Bertram, A.; *The Searchlight.*

7   Gunawardhana, C.A.; *Development of the Rubber Industry IDB Board Paper, 1971.*

8   George, Susan; *'How the other half dies' (Pelican).*

9   Gunawardhana, C.A.; *Science & the People, 1977.*

10 Gunawardhana, C.A., *Science & the People, 1977.*

11 *Proceedings, Panel on Indigenous Medicine, Ministry of Planning & Economic Affairs, 1974.*

STATEMENT S

- Technology abandoned in the past is becoming the technology of the future.

- Appropriate technology signifies different things to different people in different countries. One is aware as to what is inappropriate technology, and through a process of elimination (or selection) one can arrive at a form of technology which may be appropriate in a given situation.

- It must be realised that Sri Lanka, like most of the developing countries has not been a country isolated from the rest of the world. The choice for appropriate technology in a development programme, must be given priority due to strategic reasons.

- Appropriate technology can never be imposed on people. The system should be presented in such a manner, so that people can select and decide as to what is appropriate to their situation, and that what is not.

- Commercial viability is an essential prerequisite in deciding upon the application of a particular technology. A particular technique which may be viable in one context may not prove to be viable in another.

- Activity in the sphere of technological development cannot be influenced by a political system, be it liberal or socialist; however technological development could bear an influence on any Government, irrespective of its political background.

A UNIVERSITY PERSPECTIVE ON APPROPRIATE TECHNOLOGY

*Robert P. Morgan**

I was asked, as we all were, to outline my ideas on the general objectives of
the Workshop, which include enlarging the knowledge base about processes basic
for introduction of appropriate technology (A.T.) and evaluation of A.T. theories. I have neither been heavily involved in implementation of A.T. projects
nor I am much of a theoretician. Therefore, I will build my remarks upon my
experiences which have been almost entirely concerned with university teaching
and research and then try to tie these remarks to the Workshop objectives. A
study we recently completed on Appropriate Technology for Renewable Resource
Utilization will also be briefly described.

Let me begin by stating that I have read the literature of the Center for Appropriate Technology (CAT) and am most favorably impressed by what has been
accomplished here. The creation of CAT, in a major technological university,
is an important step forward in putting appropriate technology on a firm
footing for the long-term. Universities are essentially conservative places,
hopefully in the best sense of the word. They are sometimes slow to accept
new ideas. As one who has sought to bring about change, I can well imagine
what you have been through in establishing CAT. Particularly notable is the
involvement of established departments and sections in concrete projects directly linked to overseas development. I have had little success to date in my
own university in this regard. My guess is that your success is a combination
of the right people, hard work, luck, and a guaranteed government budget for
education and research which encourages participation in A.T.

TECHNOLOGY AND HUMAN AFFAIRS AT WASHINGTON UNIVERSITY

I left traditional university positions in chemical and nuclear engineering in
1968. My prior involvement since 1961 in Volunteers in Technical Assistance

**Washington University, Missouri, United States of America*

(VITA) had given me a somewhat non-traditional perspective on technology and a strong interest in its role in international development. It is a perspective that can be described as *needs based* and it comes about in a variety of ways. In my case, reading and attempting to answer requests for technical assistance at the village level provided insights which could not be gotten in a more conventional academic or industrial setting in the U.S. The late 1960's were a time of turbulence and change on university campuses; Washington University and its School of Engineering were receptive to change and our Center for Development Technology (CDT) (then called the International Development Technology Program) and our Department of Technology and Human Affairs (THA) were born.

For a variety of reasons, our activities have taken on a character that differs somewhat from that at CAT. First, our research and studies tend to be planning and policy oriented as opposed to hardware oriented; that does not mean that we will not do hardware research and development -- it just hasn't been a major part of our efforts to date. Second, our activities tend to be oriented primarily towards activities and problems in the United States. We do retain a strong interest in international development work but it is not a predominant thrust of what we do.* Third, some of what we do you might find inappropriate. We have done studies of innovation in the chemical industry, of benefit-cost evaluation of breast cancer detection strategies, of communication satellite applications in education, and of earth observation data management systems. We don't use the word *appropriate* in the title of our department or center. I find that many engineers and non-engineers I deal with are uncomfortable with it, although Delft's initiative in establishing CAT and other initiatives may change that. It will be interesting to see how the A.T. name and concept does and the form it takes five or ten or twenty years from now. Finally, our research activities are heavily intertwined with teaching activity, with most of it housed within a single, non-traditional department in the engineering school. The focus of our Department of Technology and Human Affairs, which offers bachelors, masters and doctoral degrees, is on technology -- its applications and implications -- and on technology-

*A recent paper which describes the international activities of our department and center in more detail is Morgan, R.P., "University Education for Technology and International Development : The Program at Washington University, St. Louis, U.S.A." Paper prepared for Segundo Simposio de Ingenieria, Technologia Apropiada para Paises Subdesarrollados, San Salvador, El Salvador, Feb. 19-23, 1979.

policy analysis. Students who take any of my courses will run into appropriate technology in one form or another; students in other courses in our department may not.

Currently, the demand for our students in government and industry is quite good. For example, seven of our graduates are now working for the U.S. Environmental Protection Agency. One graduate is working full time for VITA. It has occured to me on more than one occasion that our students are assuming positions of some importance. The kind of education they receive from us is important; our responsibility to those students is significant. I don't think we can sell them on a particular point of view but we can expose them to differing points of view -- the small-scale as well as the large-scale, the labor-intensive as well as the capital-intensive, the resource conserving as well as the resource consuming, the global view as well as the national view. There is a strong demand now for engineering education in the United States. Traditional programs such as chemical and mechanical engineering are attracting large numbers of students. At the same time, the energy situation and resource depletion are forcing a reexamination of both technology and lifestyle, of previously discarded technologies and the creation of totally new ones. What we deem as appropriate for people in other countries may very well prove appropriate for ourselves. The situation is in many ways unprecedented and presents great opportunities as well as dangers.

The task as I see it for the technological university, is to maintain strong programs in engineering and science while, at the same time, broadening the context in which engineering and science are considered and used. Thus, in our new bachelor's degree program in engineering and public policy which has substantial engineering content, we also require coursework in political science, economics and in technology and human affairs. A project or policy study is also required which, if it is done right, will provide the student with a real-world experience outside the university where the possibilities for A.T. involvement as well as planning and policy studies exist.

A.T. FOR RENEWABLE RESOURCE UTILIZATION

We recently completed a study entitled Appropriate Technology for Renewable resource Utilization as part of U.S. preparations for UNCSTD.*   In this

*Morgan, R.P., Icerman, L.J., et al., "Appropriate Technology for
Renewable Resource Utilization." Final Report submitted to Aid,
Center for Development Technology, Washington University, St. Louis,
MO. 63130, USA, June, 1979.

study, which was carried out in collaboration with VITA, information was gathered and analyzed for five technology areas of current or potential use in
developing countries : wind energy utilization; improved cookstoves; solar
drying technologies; rice bran utilization; and materials and products from
natural fibers, agricultural residues and timber wastes. We gathered information from many parts of the world from a variety of organizations. We then
proceeded to define initiatives that the U.S. government might support in
those areas. In general, there appears to be a need to:

1) collect, evaluate, and publish data on village level needs and local
   resources to help meet those needs;

2) provide information dissemination programs on appropriate
   technologies;

3) support innovative extension efforts in connection with
   these technologies;

4) perform research within and among developing countries and
   collaborative research between developing countries and U.S.
   institutions in order to improve and adapt designs;

5) carry out work as in item (4) above to test designs and
   evaluate performance of the technologies in question;

6) cooperate internationally in the evaluation, assessment, and
   publicizing of efforts to utilize specific technologies in
   developing countries.

More detailed initiatives for each of the study areas are included in the study report.
The degree to which each of these activities should be emphasized depends upon
the technology or product in question. For example, in direct biomass utilization for cooking, the literature and report of a VITA-CDT Expert Panel assembled during our study indicate that some improved cookstove designs are already available; therefore, extension efforts to disseminate those designs
should receive high priority. However, in solar crop drying and rice bran
processing, we came across few if any appropriate, low-cost designs that appear to have been widely utilized; thus considerable applied research and
development work might be needed more before extension efforts could go forward. More analysis, including field surveys, is required to define priorities
among initiatives with greater certainty.
It is our feeling that there is potential in each of the areas we examined for

bringing about improvements in living conditions among the poor in developing countries if sufficient resources and effort are expended. However, social, cultural, and economic factors as well as government policies are complex and can strongly affect the outcomes of such efforts. There is also for each technology or project area, considerable uncertainty about the cost and reliability of the technologies. Furthermore, an economic comparison of the technologies which were selected for study with other technologies was beyond the scope of our study and remains to be performed. In some areas, (for example, cookstoves) the users may be individuals whereas in others, such as windmills, a village or community effort may be required because of cost. Both of these scales were deemed appropriate in our study.

A key element in using appropriate technology for renewable resource utilization to improve the life of poor people in developing countries is to strengthen the ability of those countries to carry out the activities described in the initiatives. Developing country organizations as well as developing country individuals are the principal means for achieving this goal.*   Developing country organizations that could participate in such efforts include universities, government organizations and A.T. organizations; specific organizations are listed in our study report for each of the five topical areas. The documentation associated with the 1978 New Delhi Forum on Appropriate Industrial Technology sponsored by UNIDO indicates that in many countries, a considerable amount of activity that falls within the topical areas of our study is going on within the developing countries themselves. Much A.T. and other developing country work goes on locally and does not make its way into international information channels.

A variety of organizations in the United States and Europe have also been identified which have either been involved or have the capability of becoming involved in implementing initiatives. Although the role of these institutions is necessarily a limited and supportive one, it may be that contributions can be made, particularly in the areas of R & D and information dissemination. A number of mechanisms exist for cooperation among developed and developing countries; however, the financial support being provided by some developed

---

*There is a point of view held strongly by many A.T. organizations that unless the poor themselves are involved in the process of defining needs and choosing or developing the technologies to help meet those needs, neither the technology nor the development which takes place will be truly appropriate.

countries for such cooperation would appear to be inadequate.

We recognize that it is not possible to focus exclusively on hardware and expect a study such as ours to lead to beneficial change, since many of the areas we are talking about involve significant social, economic, political and environmental factors. On the other hand, the technologies and products which we consider appear not to have been examined in any kind of systematic way and therefore the data base -- the knowledge of what they are needed for, what is available, how well they work -- is lacking. Thus, we chose primarily to examine specific technology or product areas but we did so with some consideration of the non-technological factors which strongly govern the success or failure of innovative efforts in appropriate technology. Inter disciplinary field investigations and more detailed economic comparisons of alternative technologies are required to shed further light on the status of the technologies under study and their potential for widespread use.

The study I have briefly sketched is no substitute for the kind of grass-roots involvement that many A.T. organizations stress. However, it does attempt to pull together and analyze information on technology areas potentially useful in development. It can help to identify new R & D opportunities on one hand, and possible approaches for immediate application on the other. It can serve to bring some structure to the A.T. field. It can serve as a bridge between A.T. work in the field and the more traditional resources of the university. My own feeling is that non-traditional centers such as CAT, associated with educational institutions such as Delft Technological University can be very helpful in both developing and developed countries in bringing about this kind of bridging effect and contributing to international development.

STATEMENTS

- Especially non traditional Centers such as the Center for Appropriate Technology can be helpfull in the sense that these can bring about a kind of bridging effect between the institutions they are part of and the places their work relates to. These centers can thus significantly contribute to international development in both developing and developed countries.

- If a close and good working relationship between universities and appropriate technology groups can be established this will be beneficial for appropriate technology in the long run.

- Today there is a tendency for engineering schools to get involved in appropriate technology more and more, especially when it is described in topical terms (e.g. solar energy, bio-gaz, renewable energy sources). It is important for educational programs in the Universities to expose students to all points of view, as well as the concepts behind these.

A "COMPANY FOR RURAL DEVELOPMENT IN INDIA

K. Krishna Prasad*

I. INTRODUCTION

"If you want to go places, start from where you are. If you are poor,
start with something cheap. If you are uneducated, start with something
relatively simple. etc., etc."

This sound like the opening lines of a lyric in competition with
"If I were a rich man ..." sung by the here of a famous Broadway musical,
"The Fiddler on the roof". Sorry to be boring; it is nothing as exciting
as that. It is the quotation attributed to Schumacher (1973) placed at the
head of Chapter 1 of a recent book on appropriate technology (Dunn 1978).

It is interesting to compare the two. The first one is addressed to a
nameless "you"; the second one sings about an "I". The former is an
admonition served on a poor man by a wealthy person; the latter is
the dream of a poor man. And there in lies a fundamental contradiction
in much of the polemics that goes to make up what is called "development
science".

Let us delve a little deeper into the significance of the "you" in the
quotation. In principle the "you" is a poor villager in some far of land
like Upper Volta, India or Columbia. Unfortunately he is inaccessible.
Therefore, in practice, the message has to filter through the agency of
an intermediary. Every bit of rhetoric on development, everything said
and written about appropriate technology, and in fact every bit of gadgetry
is meant for the consumption of the intermediary.

*Eindhoven University of Technology, Eindhoven, Netherlands

Who is this intermediary? Depending upon the persuasions (or, should I say, ambitions) of the purveyor of the message the intermediary can be the nearest "development worker", the planner of one of the seventy add countries that are labelled poor, of a Un.N. agency.

Yet the message is about self help approach to development. If self help could have achieved the stated goals of development, conferences of the present type would be redundant. No, that is not true; an intermediary is obviously essential. This is how the second contradiction comes about in the well publicized thinking on appropriate technology.

That is not the whole story either. It is generally assumed that technology is the main driving force for development. The assumption is based upon the wonderful achievements that technology has brought about in Western Europe and North America. It is rarely pointed that these achievements were built on the solid foundation of one hundred and fifty years of capital accumulation as a consequence of an exploitative economic order. The industrial enterprise of those days was not conceived as a means of national development, as it is understood today. Yet technology - appropriate or otherwise - is being preached as the basis for national development. This is the third contradiction in the whole system of beliefs that con- stitutes development economics.

Of course the protagonists of appropriate technology will raise in protest at this reading of the situation. They will point out that they take into account the so-called social problems of introducing new technologies into 'rigid', 'inefficient', 'inequally structured' poor communities. That is true. But the appropriate technologists are busy filling a large "do-it yourself" store with gadgets - actually filling a large loose-leaf folder of "how to" information - without much concern about how this information is to get across to the poor people of this world.

Lest the author be misunderstood, he would like to state that he is in total sympathy with the general philosophy behind appropriate technology. What is being challenged here is the assumption that the mere development of a relevant technology will produce the desired objectives. What is not recognized with sufficient seriousness is that the conditions of rural communities demand the intervention of a well orchestrated group of intermediaries with adequate time, talent and money at their disposal.

The main thrust of the present paper is to delineate the character, structure and functions of such a group for rural development in India. The need for such an approach to development is sketched with reference to the historical process of technological development in Western Europe and North America, a short summary of development efforts in India and a view about what the approximate final 'developed' state in a reasonable time span ought to be. The treatment of the first two aspects may be dismissed as rather contentious; the short space available does not permit a detailed proof of the points raised. The third one is simply a statement of desire on the part of the author as to what should be the state of development at the end of specified period. Unless one wishes to be a prophet of doom, that is the best an excercise in futurology could hope to achieve, however sophisticated the approach adopted to arrive at the final 'scenario'.

The purpose here is to present thoughts as they have occurred to the author from time to time and not to present an erudite commentary on development processes.

II. A HISTORICAL PERSPECTIVE

The technological witardry and the liberal social organizations of Western Europe and North America in the present century have to be viewed against the background of one hundred and fifty years of capital accumulation under the influence of industrial revolution. The principal features of this phase of development are summarized below.

a) The industrial enterprise (Which is really a manifestation of technology) was not seen as a means of national development (as it is understood now), but simply as a method by which a few people could get wealthy and powerful.

b) The large scale colonization provided a multitude of benefits: direct cash, exploitable natural resources using virtually slave labour, and a ready market for the industrial produce at extortionist prices.

c) Opening of new continents – The Americas, Australasia, and Southern Africa – provided for substantial emigration from Western Europe and helped relieve pressure on local resources.

The Welfare state has banished poverty from its boundaries. Except, I am
not sure that it has not contributed to poverty elsewhere.

The capital accumulation process described above is unable to explain the
development in the technological business, is truly intricate and a thorough
discussion of this is way beyond the scope of this paper. I shall be content
with illustrating this intricacy by an example.

The development of the automobile in the early part of this century and its
near perfection in the next forty years of so is, I believe, one of the
classic examples of a continuous collaboration among various constituencies -
idea generation, actual production, marketing, using - involved in the
phenomenon of technological development. It is not surprising that for a long
time the United States of America was the headquarters of automotive innovation in the world. A naturally wealthy country with its vast size and people
with a highly individualistic attitude were looking for a fast, reliable and
safe personal transport. The industrial enterpreneur developed the product;
the consumer was willing to pay for a rather untested product and patiently
tried the system out.

Being a mechanical engineer starting work in the field of internal
combustion engines, I have never stopped at marvelling the scientific and
engineering feats of the community which produced a system that could be
operated with relative ease by virtually a technical ignoramus. Each of the
main components of an automobile - the engine, the drive assembly,
the suspension system and the tyres - required years of patient research
and development with contributions from industry, university and consumer
usage. The concept of after-sales service by specially skilled technicians
available almost at any place where an automobile is likely to go and the
taxation system that was evolved to pay for building and maintaining the
roads for these vehicles to use, made their own contribution to the popu-
larity of the automobile. The special role played by the University system
in providing sophisticated technical manpower can not be left out of the
reckoning in this process of development. The contributions to the development of the automobile was not confined to national boundaries; in spite
of fierce competition among the industries, there evolved a methodology of
exchange of technical know-how through motor shows, conferences and product
marketing all over the world.

The historical perspective has many a moral to tell. To the planner in an underdeveloped country, it says that mere adoption of a slogan like "industrialize or perish" will not result in industrial development. The conditions of capital accumulation available to Western Europe and North America during the last century are denied to an underdeveloped country. To the entrepeneur it says that the fundamental characteristic of a successful technology is that it produces a product which fulfils a widely felt need. To the appropriate technologist it points out the vital aspect of several constituencies involved in the diffusion of technology and its products. To the cost benefit analyst, it says that technology is an evolutionary process and requires vital inputs from in-service operation. The last point needs particular emphasis since many a cost benefit analysis of some recent technologies of rural relevance tends to ignore this aspect of technological progress.

III. A SHORT REVIEW OF DEVELOPMENT EFFORTS IN INDIA

The development effort is largely government inspired. Two broad categories can be identified. One is concerned with industrial development and the other is agricultural. The industrial sector - both government and private - for the most part attempts to imitate the development that has taken place in Western Europe and North America.

The government sector has been primarily responsible for the development of power generation, heavy industry, transport and communication system, the basic infrastructure of education, health and research facilities. The private sector has some minor inputs to the above, but concentrates on a spectrum of consumer products.

The agricultural development has been sought to be achieved through large scale reservoirs with associated network of canals, rural electrification programs providing for pumping from wells, fertilizer plants and considerable inputs to 'green' revolution.

Thirty years of this activity has resulted in considerable developments. A crude guess would put the population that has benefited from this enormous activity at 50 million. The condition of the rest of the population is said to have either remained stationary or further deteriorated. The whole plan-

ning activity has been heavily criticised for this lop-sided development.
Of course, in the light of previous discussion, the situation could not have
been otherwise. The capital accumulation essential for such a phenomenon to
be felt by a larger section of the population has been been materialized.
Conditions that existed in nineteenth century Europe and North America are
simply not available in the 20th century for countries like India.
It is not as though the different governing bodies have not realized this.
Considerable effort has gone into looking at the problems of the poor in
the rural areas. This activity has taken a variety of forms: land reform,
providing credit facilites, formation of co-operatives for performing a
host of functions, bringing in health and educational facilites. None of
these seem to have provided the benefits that were expected of them to the
poor. Several reasons could be cited for this state of affairs. The landed
gentry and local officials of the government form an effective clique to
prevent such benefits from reaching the poor. The poor themselves are un-
aware of most of the provisions of different legislations supposedly in
their favour, but also not organized themselves to see to it that these
benefits reach them. When they do attempt anything like that, they seem
to get involved in endless legislation out of which the poor and deprived
rarely emerge as victors.

The single agent of the government who could have provided the necessary
inputs for developing an effective organization among villagers is the
Block Development Officer. (A block may comprise of as many as 100 villages).
This institution is the brain child of the Community Development Programme
conceived in the mid-fifties. The institution suffers from innumerable
shortcomings. His function is supposedly to be in charge of community
development, but is not neccessarily in charge of all government activities
connected with rural development. There is a case of severe one-upmanship
among officials involved in the different sectors so that the effectiveness
of the different programmes is considerable attenuated. The job is not a
highly remunerative one and is expected to be carried out in an environment
which lacks even elementary public amenities. Moreover he occupies one of
the lower rungs of the bureaucratic heirarchy.

It is obviously not a job that attracts the best of talent in the country.
Hence the general feeling in the country is to identify the person occupying
the position as one who appears singularly devoid of intelligence, imagina-

tion, or initiative - qualities that are obviously required of such a person. Lastly, but not the least, he is generally part of the establishment mentioned earlier. By accident, if there happens to be a good officer, the establishment sees to it that he is moved to a job which is relatively 'harmless' of which the governments of this world seem to possess in abundance. It is not surprising under these conditions that villagers treat any person who even remotely resembles a government offical however deeply he may be committed to improve their lot with scant respect and dismiss him without as much as listening to what he has to say. Which makes the problem of rural development so much more complex than it really ought to be.

So much for government efforts at rural development. The second type of effort is that contributed by voluntary agencies. The agencies are of different varieties (for example Gandhian movement, Christian missionaries, private groups and foreign volunteers) and have had a long history in India. Most of these groups are very small (as small a single individual or husband-wife team) and tend to concentrate on providing basic amenities like health services, schools, drinking water supply etc. While their work is regarded highly successful, their smallness works against them. Because of their smallness, they are able to bring only a limited talent to the total task of rural development.

They also operate on shoe-string budgets which inhibits the real professional joining them. Since people in this group work and live in rather difficult conditions, there is a general tendency among many intellectuals to dismiss them as starry eyed individuals who will be able to achieve nothing significant. Their smallness makes them particularly vulnerable to attacks from the establishment in case they attempt to overstep the mark of unwritten law in the village. The groups that rely on volunteers from abroad are rarely able to get off the ground as it were because of the rather rapid turn-over of the personnel. A general assessment of such work would be,their impact is quite limited in comparison to the magnitude of the task - both in terms of the number of villages handled and in terms of the spectrum of services provided.

Yet another class of development effort can be noted and is primarily inspired by the social elite in large metropolitan centres, partially or completely supported by Government. This effort covers the specialized sector of handicrafts and weaving industry in rural and semi-rural areas. I have no intention of running this class of work down, as it prevents extinction of the tradition of rich artisan skills that India is heir to. The number of people involved in this type of work and the market for such products is very small. One should not make the mistake of identifying this class of work with rural development as is often done.

We end this section by noting that the need is really for talented professionals to be involved for genuine, long-lasting, rural development.

IV. A 'FUTURE FOR RURAL INDIA

Having gone into the rather unsatisfactory state of affairs at the moment, two obvious questions arise :

1) what should be the goals of development? and
2) how do we go about achieving these goals?.

It is to the first of these questions that we turn to in this section.

While developing these goals, several assumptions have been made. The basic premise is that an egalitarian society in India can not be build around European technological witardry which is highly exploitative of all kinds of resources - human, natural and environmental. Once this premise is accepted, slogans like "industrialize or perish", "trade but not aid" etc., etc., will fall by the way side. India is one of the more fortunate countries in that it possesses enough resources to maintain a more of less self-sustaining system.

Agriculture is identified as the priority sector for development. Rural industry is often touted as the basis of generating employment in the country side. This is challenged here as unrealistic. Unrealistic because questions of the following type are difficult to answer: What are the products? What is the market? It is simply useless to speak about cement from rice husk and fibres from sisal (to name just two ideas that seem to be so popular among appropriate technologists), for urban markets. These products can just not compete with similar machine-made products which have a strangle-hold over urban economy. Moreover, it is simply unthinkable of providing employment for 150 million people or so on the basis of such

industries to feed a market which can at best be generated by 50 million or
so well to do urban population. Investment in such activities is the surest
way of perpetuating poverty among the rural poor.

If we take note of the fact that technology can only flourish when it produces
a product that satisfies a widely felt need, we assume that agriculture is
the technology for India. It produces a product that satisfies a widely felt
need - food. Can agriculture be productive as well as absorb more labour?
If the Western model for agricultural development is given up, it apparently
does. Here are some facts that support this assertion (Lappé and Collins,
1977). Japan and Taiwan, both thought of agriculturally successful, have more
than twice as many agricultural workers per hectare than Phillipines and
India. The value of production per hectare is seven times that of the
Phillipines and ten times that of India. What is needed is the adoption of
labour-intensive farming techniques that *productively* uses labour.

Once the agricultural productivity shows a true upward trend, it generates
its own needs apart from the needs of people whose food requirements are met.
This is seen in the present work as the basis of rural industry. Such an
industry develops according to the needs of the people and resources base of
the environment.

The model sketched below is based upon a suggestion originally made by
Roger Revelle (1976) that it is possible to conceive of a more of less
self-sufficient ecosystem. Many numbers used in arriving at outputs are
based upon the work of Revelle, Arjun Makhijani (1976) and Reddy & Krishna
Prasad (1977). The analysis is based upon a national outlook rather than
that of a single village. There is no particular mend in using a village as
a model system, since they come in various sizes (population from 300 to
10.000 or so), from differing climatic zones and different resource and
skill situations. Averaging involved over such a diverse conglomerate does
not shed any more light than the national picture presented here. On the
contrary, the present approach clearly focusses attention on the magnitude
of the task.

Table 1 summarizes the basic data that have been used in arriving at the
schematic diagram of a 'developed' rural economy in India.[*]

The main energy source for the system is solar energy via the natural process
of photosynthesis. One hundred million hectares (a little under two-thirds

---

[*]*Detailed calculations and assumptions made in arriving at these numbers
are available from the author.*

of the land under cultivation) double cropped every year with appropriate inputs of seeds, water, fertilizer and pesticides can yield food of the order of 400 million tons per year. This number is over three times the total food output in 1976-77 in India and the per hectare output is little less than half that attributed to Japan.

The food grains always produce wastes which are fed to cattle. The cattle dung as well as unused agricultural wastes are to be used in a well-engineered Bio-gas plant operating at about 35°C and an efficiency of 60%. This efficiency is easily realizable if one examines the data provided by Parikh and Parikh (1976). The gas plant produces energy to meet all the needs of irrigation, mechanical energy for agriculture (to be thought of as a means to overcome the seasonal peak labour demand that is characteristic of agriculture in India), and partial fertilizer input necessary to raise the food indicated.

The other major energy consumer in rural areas is cooking energy. This is sought to be met out of forest resources. The area required for this is set at 20 million hectares (which is roughly 30% of the forest lands currently available in India, see Langerhorst et al, 1977). This estimate is based on rather modest assumptions: o.25% photosynthetic conversion efficiency; availability of cooking stoves that use energy at double the present value; and the population doubles during the period under consideration and taking into consideration the previous assumption the total consumption remains at the same figure.

There is excess energy available from the gasplants left over after meeting the agricultural energy. This could serve several purposes :

a) part of it could be delivered to urban industrial sector as a compensation to energy siphoned out from there for fertilizer production;

b) part of it could be sold to customers in villages who would like to use gas for cooking; and

c) part of it could be used to drive the engines of rural industry.

Note that if demand b) goes on increasing, it could be met by using the fuel wood from the forests in pyrolytic conversion plants to produce gaseous fuel. By and large, the system is self supporting.

The gas plant is suggested to be operated with a compressor-cylinder
arrangement. It provides for greater flexibility in that gas can be supplied
to the point where it is needed and can adjust more rapidly to changing
demand patterns, as against a piping and blower system. Secondly, the waste
heat provided from the engine-compressor system can meet a substantial part
of the heating requirement of the fermentation reactor; the rest is to be
met by solar energy or burning part of the gas generated. Moreover the amount
of steel required for the gas plant of current designs could be cut down.
From the cost figures available to the author, it appears quite competitive
with the piping system.

It is not the intention here to assert that the bio-gas plants are the only
means to bring about the developments sketched here. Pyrolytic
conversion plants which are claimed to have efficiencies of the order of
85% or so could be considered as an effective alternative, provided the
agricultural waste involved has no fertilizer potential. If the latter is
the situation, then the biogas plant is the preferred system. It is also
not the intention to suggest that wind energy systems and direct conversion
of solar energy systems have no use in the final state of the economy
vizualized above. What is being asserted is : the agricultural activity
produces waste, which has energy and fertilizer potential, which has to be
exploited first before looking at other alternatives.

Some cost numbers are depicted in Table 2. It seems quite competitive with
coal based electricity system. This probably is one of the few renewable
energy resources which need not await cost escalation of fossil fuels for it
to become competitive. In addition the energy system provides jobs in rural
areas rather than in some far off coal mines or in a central electricity
station. The whole system needs additional investments by way of wells,
pumpsets, mechanization equipment etc., etc. This has been estimated to be
roughly double that of energy production system, or a total investment of
Rs 10.000 per hectare of land. It can be expected with the correct choice of
technology, the whole proposed should generate jobs on a full time basis for
at least double the present labour force. If the scheme has to have any visible impact, a 5% growth rate should be achieved so that the results could be
seen in the life span of many a thirty year old. The actual period will work
out to seven five year plans. The fifth five year plan had an outlay of about
Rs. 650 billion. If the outlay is assured to remain at the same level in

constant rupees over the period envisaged, the investment in the scheme
turns out to be just under 22% of the total outlay. This leaves a handsome
margin for the industrial sector to meet the needs like electric-power, steel
engines, compressors, pumps, etc., etc. - and industrial development that is
a response to genuine needs of people.

V. THE 'COMPANY'

It is expedient to break up the rather ambitious scheme outlined in the
previous section into smaller parts. However the small part should be large
enough to be representative of the whole. Such a representative size would
cover about 10.000 villages - roughly 2% of the total number of villages in
the country.

This villages are expected to have 3.2 million hectares of land under
cultivation. About half of these are said to be owned by 2% of rich families.
The initial scheme is to cover only the remaining half and according to the
2/3 formula adopted in the previous section, 1 million hectares of land are
to come under its jurisidiction. Care should be excercised to choose the
land from among farmers who own really small chunks of land. Answers to
questions like "What happens is such lands are not contiguous since digging
wells on ultra-small pieces of land will be uneconomical?" have to be found
as part of the execution of the scheme.

The budget for this pilot scheme would be Rs. 10 billion. A minimum period
of 10 years will be required to develop the methodology required to achieve
the goals of the scheme. During the last years of this period enough will be
known about the methods and progress achieved so that steps to replicate the
pilot scheme to cover the entire country side could be initiated. The reason
for the rather enormous size of this pilot scheme will be developed in the
discussion that follows, which will exclusively devote its attention to the
pilot scheme.

We now turn to the second question posed at the beginning of last section.
Even the pilot scheme is quite large and can not be put through without the
active involvement of a solid case of well-motivated intermediaries with
considerable professional talent. The types of organizations discussed in
section 3 cannot be expected to attract this class of professional talent on

112

the scale required. Who are these professionals? Can we find such talent in India? How can we attract such talent? What will be their organizational forms? What will be their functions? How is the system to be financed? The purpose of the rest of this section is to provide a plausible set of answers to these questions.

Since the scheme involves a whole set of functions (some of which have not been explicitly treated in the previous section), it is expected that there will be need for every conceivable professional - agronomists, doctors, economists, engineers, lawyers, management people, scientists, small scale industrial entrepreneurs, sociologists, teachers, etc., etc.. During the past thirty years, there is a considerable talent built up in the country - a talent that was a natural consequence of the development programmes in different sectors. Though such programmes were not of benefit to the poor, this talent can be put to use in the scheme envisaged here with profit.

A substantial number of these people are quite competent. They have also given serious thoughts to the state of affairs in the country and would like to do something truly useful. However, it is not possible to attract such talent on the basis of the schoolmasterish admonition that we began this paper with. Neither would they be willing to live like a Ghandhi. Three conditions seem essential for attracting such talent :

(a)   the task should be challenging;

(b)   an organizational form which permits the use of their talent without extraneous interference; and

(c)   their creature comforts, requirements of job security and fulfilment of family obligations should be met to a reasonable level.

That the task is challenging enough is easy to see. Covering something of the order of 10.000 villages implies an area of approximately a third of a state like Karnataka. The fact that the poorest third of the farmers is chosen posed a formidable problem of trying to convincing these into forming themselves into more viable units so that the benefits of irrigation could more easily reach them.

It is the firm belief of the present author that the initial group that
starts work on the pilot scheme should comprise of 40 to 50 people of 20 - 25
family units. They should be a cohesive group with approximately similar
ideas about the work involved and preferable possessing similar life-styles.
The group will be called *the Company* in the rest of the work, for lack of a
better name. The company even at this stage is large enough: to house the
different types of talent mentioned earlier; to be able to withstand the
'static' from the local establishment; and to withstand the inevitable
attrition effect of people dropping out. The company acts as an autonomous
body rather than a wing of a Government Department and as such will have
adequate manoeuvrability.

The company will *not* establish its base in the nearest large town from
which it will operate. On the contrary, it will dig its heels in a rural
area. It will acquire about 10-15 hectares of land which will provide for
housing and a reasonable amount of agricultural acitivities, with room for
moderate expansion. The housing will not be of urban vareity with its pre-
ponderance of cement, steel and aluminium, but obviously has to cater for
space requirements of the type of personnel who are to live in them.
Reasonable facilities for water supply and sanitation is to be made
available. It will have a dormitory, office space, a meeting place, a
school and a clinic.

It is important to realize that the company colony will be rather an
isolated place and as such women from urban environments will find it
hard to occupy their time in a pleasant manner. It is essential that
they be a part of the company and as such have a professional stake in its
goals. Some of the women may find it inconvenient to work full time, but
part time work should be permitted. This will remove an important source
of attrition among the personnel.

The personel need to be paid according to the usual norms employed for
similar functions in an urban environment. Because of the relative
unpopularity or, shall we say, perceived sense of glamourlessness, of the
job, special service conditions, hiring and firing practices may be necessary. These are best left to the initial group of the company to formulate
subject to revision as the company grows. The company of course is expected
to be accountable since public funds are involved. However in the name of

accountability, it should not be put into a straitjacket. This will not only divert the company's attention from its main task but also can be responsible for considerable attriction.

We now turn to the functions of the company. The list of professionals we included earlier clearly indicates the type of functions the company is expected to perform. There is nothing new in these; the main departure of the present proposal is that these services - no doubt important - are seen as adjuncts to the main task of organizing farmers to grow more food utilizing the available resources to the best advantage. It is to this task we shall address ourselves next.

The first step is to start talking to the small farmers trying to explore the possibility of grouping them into *small* associations.** The word small is the operative one. The sentiment involved in owning land and pride in making decisions about farming operations appear to be a very strong urge among small farmers. Large associations tend to deprive the small farmer of this sentiment and pride. In addition management of such associations tends to be invested with the rich and influential in the village - a situation which the company would like to disturb. Suspicion on these courts is suffi-cient to make the whole strategy a non-starter.

Worse, the company to overcome these obstacles may be tempted to take over the responsibility of running these associations and in turn running the totality of farming operations. Such a situation is nothing but half-a-dozen petty landlords being replaced by a Super Landlord - The Company. Under such circumstances the company's energies will get sapped in the daily routine of agricultural business and may never get to the 10.000 Village according to the time schedule. Hence, it may be constrained to develop an elaborate bureaucratic apparatus to achieve the task. Once this happens all the talk about committed professionals, autonomy for the company, special procedures for service regulation etc. will inevitably have to go out of the window. Either the company will have come round full circle - back to square one -, or will become a soulless machine, the dream of many a trans-national agribusiness company.

Such procedures should be denied to the Company by charter. It *has to spend* time talking to the farmers, device special examples, and invent 'games'

** *The word 'association' is used rather than 'co-operative' as there are apparently difficulties of the latter not being exclusive. Our associa-tions are exclusive in the sense their membership is open only to small farmers owning land in contiguous areas!*

illustrating the futility of the present practices and advantages of working
as a small association. It can draw up draft articles of association for
discussion with the farmers. These can be changed to suit the farmers' needs.
Of course the company, in addition, can offer a number of incentives. It can
arrange for: credit for sinking a well; technical expertise for the choise
of a well site; putting a pumpset; 'working capital' for better seeds and
fertilizer. It can insure against losses (due to some unforeseen circumstances) when new techniques of farming operations are adopted. It will run demonstration farms on its land illustrating the possibility of utilizing inten-
sive-labour techniques which provide higher yield.

The formation of these associations should be viewed as the principal task
of the Company - probably the most difficult task. It will demand a great
deal of patience, tact and ingenuity on the part of the company to pull this
task through. Every member of the group should contribute to the campaign.
The doctor treating a patient gently suggests to him to consider becoming a
member of one such association. The nutrition expert talking to the woman
about how to improve diet for children puts in a word on behalf of the scheme.
The agronomist, who is talking to a small farmer about the inter-row planting
of a second crop, sings the praise of the association.

The first twenty villages would present the most formidable task - everything
is so new and there is no 'experimental' result to back the 'theory'. The
difficulty is compounded by the fact that the campaigners themselves are
inexperienced with rural altitudes and will have difficulties in separating
the rational from irrational objections to the procedure.
Once the associations get going, the problem of improving agricultural operation to attain the necessary high yields can proceed with relative ease.
Not that this will not face resistance; but it is a technological question
which presumably can be answered through well chosen experiments on the company farm.

The next step is for the associations to seek a more active voice in councils
at the village level and this is the stage at which the company will start
getting flak from the establishment. The probability is that it will start
much sooner. The thesis here is that the Company is uniquely endowed to
wealther the assault from the establishment. Professionals of the type who
are members of the company have a tendency to know people in right places.

One fellow went to school with a chap who is a deputy secretary in
the department of home in the State Government. A second fellows's uncle
happens to be the Chief Engineer for irrigation projects in the State.
A third fellow used to be the neighbour of a kid (who was considered a rebel
in school) now happens to be a member of the State Legislative assembly.
These 'contacts' can be made use of by the company to considerably alternate
the influence of the government officials at the village level.

These contacts also provide the company with tools which were formerly the
exclusive monopoly of the establishments. These factors erode some of the
powers of the establishments. The political question is probably more
difficult to handle. If the company maintains a rather low profile in the
initial periods, it might be possible to hold the confrontation with the
establishment to modest proportions. Once the associations start throwing
their weight in village councils, intelligent politicians will realize that
there are more votes with the associations than with the establishment.
Land reforms etc. can be pushed with more vigour under these conditions. The
energy system can start functioning as soon as the associations have taken
some root.

The other functions of the company move with greater pace from this stage on.
Many studies are available now how to provide different types of services to
suit different local conditions. There is no need to discuss these here.
The next point to be discussed is the methodology by which the company can
expand its operations to cover 10.000 villages. The target has been
deliberately chosen to force the initial group to think about expansion
processes. One way would be to bring in young graduates and put them
through the paces of company's work for a couple of years so that it would
be possible to start a branch of the company at another place. It is
presumed that it would be possible to attract competent graduates in a manner
similar to the original 40 were attracted. It may also be possible to
persuade other mature professionals to join the company if the experience of
the company proves promising in the first two years. A good scheme would
put a mixture of young and mature talent in the branch. A third approach
could be to attract college students to spend their holidays working for
the company on a stipend. By these procedures it is possible to conceive
succession of branches of the company to cover 10.000 villages. It is
estimated that it should be possible to generate at least 7 branches in

about 8 years.

The company's budget for the period of 10 years is 10 billion rupees. The cost of salaries of full-time employees, trainees, stipendiaries and costs of branch development would be somewhat smaller than 1%. Some of these costs are recoverable by way of house rents, service charges for water, and fuel, etc.

A last question concerns the financing of the Company. It is obvious that money on the scale we are thinking can not be raised from private sources. Apart from this development is a national task and not a money-making enterprise. Of course this does not mean money is to be poured into the drain; it simply means that the levels of profit that can be obtained and time scale over which they can be realized are not compatible with the general philosophy of private enterprise. It has to come from the Government with a suitable initial capital – something of the order of 10% of the total budget of the scheme. The rest has to come through a variety of financial institutions that exist in the country. To give an example :
The government provides a subsidy for setting up a bio-gas plant and the remaining money can be raised easily through a bank loan. The major banks in India are nationalized and a few of these have made considerable progress in extending rural credit. Credit for construction of wells and installation of pump sets is already available. With the presence of the company and formation of the associations, such credit should be more easily accessible to small farmers.

VI.   CONCLUDING REMARKS

The main line of argument in the present paper can be summarized as follows. Conditions of this century does not permit a country like India to follow the path of Western Europe and North America towards development. Since 80% of the population live in rural areas and 80% of these derive their sustenance from agriculture, development is taken to mean increased agricultural productivity. Such productivity increases should be achieved through labour-intensive techniques rather than energy-intensive techniques. Relying on the latter in the present world situation with respect to energy will condemn a large proportion of rural poor to live on a perpetual starvation diet. A more or less self-sufficient rural ecosystem is proposed as a model for development.

Certain basic concepts behind the formation of an agency of intermediaries have been formulated. The agency - called the Company here - is simply a means by which a systematic transfer of talent from urban areas to rural areas for purposes of development is brought about. The talent backed by money, time and suitable organizational forms will produce the desired changes in rural societies to bring about development.

Several issues have been deliberately left out of consideration here. These are problems like population control, schooling, choice of rural industries, adult literacy, etc., etc. The company has the wherewithalls to provide these services in abundant measure. The company will start functioning in these areas from the very beginning. However, it is the firm thesis of the paper, that these efforts will bear fruit only in an environment of confidence among the people who advocate these and the people who are to benefit from these plans. Such an environment could be created only when the poor villager experiences the practical benefit of development - more food to eat.

The company and the farmers'associations will provide the raison d'être for the appropriate technology groups. It is the firm contention of the author that appropriate technology can not bring about the structural changes in rural society that is essential for development. Once these changes are initiated, for these changes to gather momentum, it is essential that these receive the support of appropriate technology research. Without such support, it is quite conceivable that the changes would not last long. However, it can not be denied that the appropriate technology movement has been to a very large measure responsible for many a scientist or an engineer thinking about the rural poor.

I should like to end this paper by a quotation from "Asian Drama" by Gunnar Myrdal (1971) :

"But the leading figures in this drama are the people of South Asia themselves, and above all their educated class. The participation of outsiders through research, provision of financial aid and other means is a sideshow of rather small importance to the final outcome."

(The emphasis is mine).

119

REFERENCES

*Dunn, P.D. 1978 - Appropriate Technology : Technology with a human
face, Macmillan, London.*

*Langerhorst et al 1977 - Solar Energy : Report on a study of the dif-
ficulties involved in applying solar energy in developing countries,
prepared for the Minister for Development Co-operation, Ministry of
Foreign Affairs of the Netherlands.*

*Lappè, F.M. & Collings, J 1977 - Food First, Houghton Mifflin Company
Boston.*

*Makhijani, A. 1976 - Energy and Agriculture in the third world. Ballin-
ger, Boston.*

*Myrdal, G. 1977 - Asian Drama (Abridged version prepared by Seth S.
King). Penquin books, England.*

*Parikh, J.K. & Parikh, K.S. 1976 - Potential of Bio-gas plants and How
to realize it, Paper presented at the BMFT-UNIRAR Seminar on "Microbial
Energy Conversion" at the Institute of Microbiology, Götingen, West
Germany, October 1976.*

*Reddy, A.K.N. & Krishna Prasad, K. 1977 - Technological Alternatives and
the Indian Energy Crisis. Economic and Political Weekly, Vol. 12,
p. 1465.*

Revelle, R. 1976 - Energy Use in Rural India, Science, vol. 192, p. 969.

Schumacher, E.F. 1974 - Small is Beautiful. Sphere books, London.

TABLE 1

SOME BASIC DEVELOPMENT DATA FOR INDIA

---

			millions
Population:	total		549
(1971)		urban	110
		rural	439
		rural labour force	151
		small land holders/	
		tenants	76.8
		landless labour	45.5
		others	28.7

LAND USAGE IN INDIA (1972)

		million hectares
	crops	164
	fallow	21
	pasture land	13
	forests	65
	others	65

FOOD OUTPUT IN 1976-1977 : 126 m tons

NUMBER OF VILLAGES IN INDIA : 575,000

NUMBER OF CITIES IN INDIA : 6,332

TABLE 2

DEVELOPMENT COST ESTIMATES

A. ENERGY SYSTEM COSTS

Plant costs .... Rs. 147×10^9
Distribution costs. Rs. 118×10^9
Costs of Addl.
Fertilizer plants. Rs. 76×10^9

TOTAL ....Rs 341×10^9

B. COSTS OF AN EQUIVALENT COAL

Based electricity & Fertilizer system ....Rs 607×10^9

C. AGRICULTURAL DEVELOPMENT

Costs for Wells, Pump-sets
Mechanication equipment, etc., etc. ....Rs 659×10^9

GRAND TOTAL OF INVESTMENT
FOR PROPOSAL IN FIG; 1 (A + C) ....Rs 1000×10^9

EDUCATIONAL, SCIENTIFIC & TECHNOLOGICAL INSTITUTIONS IN DEVELOPING COUNTRIES

*K.S. Jagadish, Rama Prasad and Amulya Kumar N. Reddy**

The Indian Institute of Science has displayed many characteristics typical
of educational, scientific and technological institutions in India and most
developing countries (Appendix I). Such institutions are overwhelmingly
manned by an elite with a "westernized" life-style and aspirations, often
accentuated by foreign training. This elite is alienated from the rural poor
and from their traditional technologies. This alienation is amplified by the
strong linkages with the corresponding "western" institutions. Hence, indigenous institutions derive from western institutions the patterns of organization, emerging areas of research, their trends and fashions, the sustaining
stream of ideas, techniques and inspiration, the criteria of excellence and
the sources of recognition, awards and kudos. However, completely unlike
their "western" counterparts, developing country institutions are invariably
deserted by indigenous industry which favour the attractions of imported
technologies.

The result is that educational, scientific and technological institutions
in the developing countries tend to be elitist, alienated, unwanted and
without native roots. As a consequence, they tend to be pre-occupied with
irrelevant work or with work which is only relevant to urban industries/
problems and/or to the "west".

ORIGIN OF ASTRA

A growing awareness of this perhaps unconscious boycott of the needs of
rural areas, and in particular of the needs of the rural poor, led a
group of faculty at the Indian Institute of Science to propose to the
Institute the formation of a *cell for the Application of Science and*

**A.S.T.R.A.*

Technology to Rural Areas with the acronym ASTRA meaning "weapon" in
Sanskrit. ASTRA was formally created in August 1974 as a voluntary group
to initiate, catalyse, sustain and grow in the Institute work of relevance to rural development. ASTRA, therefore, was born with a well-defined mission, viz., the generation and diffusion of technologies appropriate for rural development, and the promotion of the sciences underlying these technologies.

BASIC APPROACH OF ASTRA

To achieve its mission, ASTRA formulated an innovative approach which is
based upon its view of rural development as a socio-economic process
directed towards :

a. satisfaction of the basic needs of the rural population, starting with the needs of the rural poor, in order to reduce
   inequalities between different sections of rural society and
   between rural and urban areas;

b. increasing rural participation and control in order to promote
   the self-reliance (as distinct form self-sufficiency) of villages;

c. harmony with the rural environment to ensure the sustainment
   of development.

The crux of this approach is *a formal commitment to a neighbouring rural
area*. There are three necessary components to this commitment :

1. becoming aware of the problems of the people in that area, and in
   particular of the poorest sections;

2. making the generation of technologies appropriate to the development
   of that area a part of the academic programme of the Institute, and
   not an extra-curricular activity of a few scientists/engineers with
   an urge for social work;

3. promoting the diffusion of such appropriate technologies.

Special mechanisms were also envisaged for these components.

An awareness of the real problems of the people cannot be acquired from
the isolation of the Institute; it must be actively pursued by identifying

the felt needs of rural people through direct interaction with them. It is not a question of deciding from the alienated viewpoint of the Institute what should be the problems of the people; it is a process of learning from the people what they feel pressing problems are.

Similarly, the selection of technologies appropriate to rural development cannot be left to "wise" men in ivory towers; it must be the culmination of a process of interaction with the prospective beneficiaries.
This process should consist of the following steps :

a. generation of a range of technological options for the satisfaction of the identified felt needs;

b. presentation of these options to the people;

c. enlisting the active participation of the people in the selection of that option/those options most appropriate to their development.

Thus, the process consists of dispelling ignorance through dynamic interaction with the people, i.e., the discovery of appropriateness through flows of information, suggestions, criticisms, etc., from the people and to the people. The attitude is one of "scientists don't know which technologies are appropriate, but there is a process which enables them to find out". All this is radically different from the usual approach in rural development circles, viz., '"we know what technologies are good for them" '.

The generation of technologies must be distinguished from the diffusion of technologies - whereas the entire responsibility for generation can be assumed by an institution of education, science and technology, the diffusion of technology necessarily requires the collaboration of seve-ral types of institutions (people's organizations, development agencies, raw materials procurement and product marketing organizations, financial institutions, etc.). ASTRA's approach, therefore, is to concentrate on technology generation, and with regard to large-scale technology diffu-sion to promote it by participation in multi-institutional teams.

IMPLEMENTATION OF ASTRA APPROACH

In implementing its approach, ASTRA has focussed on two major lines of activity.

The first involves the establishment and operation of an *extension centre* amidst a cluster of villages near Bangalore:

a. to acquire a grass-roots understanding of the felt needs of the people through direct interaction with them;

b. to expose the people to a wider range of technological options instead of the two-option Hobsons's choice with which they are currently confronted, i.d., either sub-optimal traditional technologies or far-too-expensive "modern" technologies;

c. to understand technology diffusion problems by microdiffusion in the cluster of villages.

Incidentally, the Extension Centre is also intended to provide Institute faculty and students with simple basic facilities for living and working in a rural environment.

The second line of ASTRA activity involves the introduction of projects, studies and researches of rural relevance into the academic programme of the Institute, i.e., into its B.E. and M.E. projects and courses, its Ph.D. and faculty research, and its sponsored projects.

The resulting mission of ASTRA in the Institute, the role of the Extension Centre, and the linkages with the departments of the Institute, with funding agencies and with technology diffusion organizations are all shown in Figure 1.

ASTRA PROJECTS AT THE INSTITUTE

Thus far, ASTRA projects are in the areas of Energy, Building, Water, Agro-processing, Resources, Transport. The choice of projects has been the result of a compromise between the interests of Institute faculty members and the felt needs of people in the cluster of villages around the Extension Centre. In cases of mismatch between faculty interests and people's needs, the separation of technology generation work on the Institute campus at Bangalore and technology demonstration at the Extension Centre minimizes unwelcome consequences.

Energy

The energy projects fall into three sub-categories :

a. Rural Energy Planning,

b. Alternative Energy Sources,

c. Energy Devices and Applications.

Under Rural Energy Planning come the projects on rural electrification, rural energy consumption patterns, scope for wind energy utilization, biogas-based energy centres for villages and the assessment of various energy sources. The projects on Alternative Energy Sources are on low-cost windmills, design and optimization of biogas plants, biogas-fuelled spark-ignition engines, methane enrichment of biogas by carbon dioxide separation, and energy forests. On the various energy devices and applications being studied/developed are biogas burners, biogas melting furnaces, village "chulas" (cooking stoves), solar thermal conditioning and water heaters for sericulture, and wickless kerosene stoves,

Building

Four sub-categories of building projects are in progress :

a. System studies,

b. Walls,

c. Roofs, and

d. Construction.

The System Studies consist of a survey of rural housing, the development of a systems approach to rural building technology, a study of the role of transport in building activity and determinations of the energy content of building materials. The projects on Walls involve investigations of rammed earth, soil cement blocks and compacted mud blocks. Ferrocement and bamboo-polythene-lime surki roofs are being investigated. Under the category of Construction activities come the projects on the construction of the biogas laboratory at the Institute and the dormitory and faculty house at the Extension Centre.

Water

Water Harvesting and Water Lifting are the two basic sub-categories of the projects related to water. The Water Harvesting projects involve studies on rainwater collection from roofs and farm-ponds, and the Waterlifting projects are on scooter-tyre water pumps, modification of handpumps, fluidyne pumps and a survey of water-lifting devices.

Agro-processing

The agro-processing projects deal with decentralized processing of agricultural products and residues. They include the production of cellulose fibre from groundnut shells, edible cellulose from rice husk, plastics and plasticisers from castor oil, sodium silicate from rice husk, soap on a small-scale, and calcium carbonate from biogas.

Resources

The projects on resources involve studies of bamboo resources in Karnataka and of villages as eco-systems.

Transport

The projects on rural transportation consist of studies on bullock carts and on pedal-powered vehicles.

UNGRA EXTENSION CENTRE

An Extension Centre has been established 113 kms from Bangalore on the outskirts of Ungra village in Kunigal Taluk, Tumkur District, Karnataka State. 55 acres of government land which was once used for a seed farm and then for a horticultural training school, have been secured from the government on long lease. The three buildings, which were constructed for the seed farm, have been repaired and brought back into use as accommodation for Institute staff and students engaged in projects. The Extension Centre has access to about 15 villages within a 3 km radius.

The work at the Ungra Extension Centre can be classified into five categories : Natural Resource Surveys, Energy Studies, Agriculture, Water and Building.

Natural Resource Surveys

Surveys of flora and fauna in Ungra, and of the land and soils in the Extension Centre have been carried out. An aerial remote sensing survey has also been done. A meteorological station has been established and data is being collected regularly.

Energy Studies

A detailed survey of rural energy consumption patterns has been carried out for Yedavani, Pallerayanhalli, Ungra and Kagganahalli in the first phase, and for Ungra, Pura, Suggenhalli, Kilara, Arjunahalli and Hanchipura in the second phase. A biogas-based energy centre for Pura is being designed. An energy forest is being grown on part of the Extension Centre land. A study has been made of energy consumption in a brick clamp at Ungra. A survey of "chulas" at Pura has been carried out. A windmill has been installed at the Extension Centre and its performance is under study.

Agriculture

Crop production on a limited scale has been taken up. In addition, experiments on programmed vegetable gardening are in progress both in the Extension Centre and in the Pallerayanhalli school.

Water

Studies on water harvesting have involved the construction of a minor tank (100 ft long x 6 ft high with 50.000 cft capacity) and of farm ponds. Water lifting projects include field testing and studies on modified handpumps and on scooter-tyre pumps.

Building

Alternative building technologies have been tried out in the construction of a dormitory of 1540 sq.ft. plinth area and a faculty house of 484 sq.ft. plinth area.

INTERACTION WITH VILLAGERS

Interaction with the people from the neighbouring villages has taken place through

1. the energy surveys,

2. the discussions on the community biogas plant proposed for Pura,

3. the enquiries regarding the ASTRA windmill,

4. the enquiries about the ram for compacting mud-blocks,

5. the vegetable gardening experiment carried out in the Pallerayanhalli school,

6. the development of "chulas",

7. the training of village youth as technicians, and

8. the visitors to the Extension Centre.

TECHNOLOGY DIFFUSION

The Ungra Extension Centre has provided a grass-roots base for the understanding of technology diffusion into the cluster of neighbouring villages. This is micro-diffusion. To achieve diffusion on a larger-scale, ASTRA has envisaged other mechanisms :

1. meso-diffusion through the Karnataka State Council for Science and Technology whereby proven technologies, e.g., modifications to handpumps, are disseminated by government development agencies to all parts of Karnataka,

2. macro-diffusion through national organizations, e.g., All-India Handicrafts Board,

3. bilateral diffusion by linkages with voluntary group and with likeminded institutions in other parts of India, and

4. educational programmes to be organized by ASTRA, e.g., the two one-day seminars on Handpumps for about 300 government engineers from the Department of Public Health Engineering of Karnataka.

Model for the Institutional Generation of Appropriate Rural Technologies

The four-year experience of ASTRA has amply validated the model adopted
by ASTRA in 1974 for the institutional generation of technologies appropriate for rural development. This model consisting of

a. the establishment of extension centres in rural areas, and

b. the incorporation of a commitment to those rural areas into the work
   programme of the institution, is rapidly spreading.

For instance, the Development Technology Centre, Institute of Technology,
Bandung, Indonesia, the Asian Institute of Technology, Bangkok, Thailand, and
the Korea Institute of Science and Technology, Seoul, Korea, are using a
similar approach.

Methodology of Selecting Appropriate Rural Technologies

The interaction with the rural population through the Extension Centre
has lent considerable support to ASTRA's three-step methodology for se-
lecting appropriate rural technologies, viz.,

1. identification of the felt needs of people through direct contact
   with them,

2. presentation to them of a wider range of technological options,

3. a selection by the people of the appropriate option.

Role of scientists/engineers qua scientists/engineers

The above methodology has clarified what role scientists/engineers should
play as scientists/engineers. They should translate the felt needs of
people into technical problems, and then carry out the research and development work to widen the range of technical options which can satisfy the felt
need.

Transformation of Traditional Technologies

The conventional approach to generating appropriate rural technologies

131

is to start with expensive urban/large-scale technologies (most of which have originated from the West) and cheapen/simplify/down-scale them for rural use. In addition, ASTRA has found that the transformation of traditional technologies is as, or more, promising an approach. That is, traditional technologies (which may have become sub-optimal because of changed conditions, constraints and possibilities) can become a rich source of appropriate technologies if their rational scientific basis is understood and they are qualitatively improved with marginally costlier inputs of modern science and engineering.

Heuristic for R & D Perspectives

ASTRA's concept of widening the range of technical options available to the people has led to a powerful methodology for discovering R & D perspectives. If, for instance, a rural house is viewed as a system with the foundation, walls, roof and services as system components, then, for each component, there are a number of options. Further, various integrations of component options result in various system options with increasing performance efficiencies at incremental cost increases. It is from such a range of system options that the people can meaningfully select appropriate options. Accepting such an approach, the generation of an R & D perspective (and therefore programme) imply involves

1. R & D on all available and new component options, and

2. R & D on all system options.

Appropriate Rural Technology / Primitive, Low Technology

ASTRA's experience demonstrates that simple products/processes invariably require sophisticated thinking/research/development and that, therefore, appropriate rural technology is advanced high technology (and not primitive low technology) if the advanced/"high" character of a technology is to be judged by the extent of modern scientific and engineering thinking and not by the trivial criterion of scale of production or "western" origin.

Appropriate Rural Technology requires greater emphasis on Basic Research

indigenous re-development of urban/large-scale/centralized technologies can be based on pursuing the path followed by the "West". Since, however, there is no beaten track for appropriate technologies, there is no alternative to starting from fundamentals. This experience of ASTRA implies that the base of fundamentals must be stronger to generate rural/small-scale/decentralized technologies than to imitate/adapt urban/large-scale/centralized technologies. Of even greater importance is the lesson that efforts to generate appropriate rural technologies necessarily require more, not less, simultaneous emphasis on basic research and fundamental science.

Model for Generating Technical Capability

ASTRA's strategy of commitment to the problems of people in a specific rural area has led to an intense learning process through which technical capability is acquired rapidly, effectively and cheaply with existing infrastructure. The environment is the best teacher provided that there is commitment to that environment. This strategy is radically different from the conventional approach of building technical capability by identifying frontier areas from a study of the output of western institutions and competing with the west in those areas - the implicit belief here is that the West is the best teacher.

Appropriate Rural Technology - an Exciting Challenge

ASTRA's projects have turned out to be as scientifically sophisticated and technically challenging as any other work in the institute.

Approach to the Promotion of Scientific Temper

The strong interaction with the people in the selection and generation of appropriate rural technologies leads to a widening participation in innovation. It is this democratization of innovation which seems a far more promising method of promoting a scientific temper amongst the people than the usual approach of science popularization through lectures and exhibitions.

*Generation of Appropriate Rural Technologies - a Powerful Method of
Conscientization of Scientists*

ASTRA's experience has shown that the process of identifying the felt
needs of people and the presentation to them of a range of technical options is a powerful method of conscientizing scientists and engineers
into an understanding of the facts and origins of rural poverty.

Approach to Inter-disciplinary Programmes

ASTRA has become a genuine inter-disciplinary programme involving about
35 faculty members from 6 departments. This is mainly due to its mission,
which is well-defined and significant. It is this mission which has
attracted scientists and engineers,drawn them together, influenced them
and made their capabilities reinforce each other. It is also the mission
which has made the whole become more than the sum of the parts.

FUTURE WORK

R & D Programmes

It is envisaged that the future R & D work of ASTRA will involve

a. intensification of efforts in the current areas of interest, as well as

b. expansion into new areas.

The main emphasis in the current areas of interest will be the development
of an R & D perspective within each area, viz., energy, building, water, agro-
processing, resources and transport, so that work can be initiated on the
gaps in the perspective, and also the separate projects can be coordinated
into a coherent programme. Since the present thrust of ASTRA reveals several areas crucial to rural development in which no work at all is being
carried out, it is envisaged that efforts will be mounted in these areas,
viz.m water management in agriculture, agricultural implements and machinery,
health, education etc. A Health Centre and a school at the Extension Centre
are being planned.

Block-level Planning

The elaboration of these R & D programmes will be carried out in the context

of attempts to elaborate a block-level plan in which resource inputs
are matched to development needs in the Extension Centre region.
An integral part of this effort will be an attempt to understand the
basis for the self-reliance of villages, clusters of villages and blocks
as eco-systems.

Information

The mounting volume of literature on appropriate rural technologies
which is constantly accumulating with ASTRA will be organized into an
information system. It is intended that this information system will
become a service to the large number of institutions and groups which
are engaged in generating and diffusing rural technologies. At the same
time, efforts will be made to disseminate information on appropriate
rural technologies developed by ASTRA and other institutions/groups.
Links will also be forged with appropriate technology information systems
abroad.

Teaching Programmes

With the growth of understanding and expertise in various scientific and
technological aspects of rural development, efforts will be made to contribute to and develop teaching programmes in the Institute. These teaching efforts will be grown organically from seminars to workshops to
short-term courses to full courses to academic programmes leading to a
diploma/degree. It is envisaged that the products of these teaching efforts will go to development and planning agencies, District Industrial
Centres, etc.

Expansion of ASTRA

Thus far ASTRA has developed only by attracting the interests and commitment of the existing faculty of the Institute. It has not acquired
any faculty specifically appointed for the ASTRA programme. This process
of transforming the interests of existing faculty is approaching satura-
tion. Hence, the implementation of future programmes required the re-
cruitment of faculty specifically dedicated to the ASTRA programme, along
with support staff and more intense student involvement. In short, now

that ASTRA has demonstrated its viability and significance, a major expansion of ASTRA both at the Institute and at the Extension Centre is
under consideration.

STATEMENTS

- In the concept of appropriate technology the word appropriate must be
interpreted as "appropriate to development". In this sentence development must be defined as a process of socio-economic change directed
towards :

  1. satisfaction of basic human needs

  2. social participation and control

  3. ecological soundness.

- The selection of technologies appropriate to rural development must,
as the culmination of a process of interaction with prospective beneficiaries, consist of three steps :

  1. generation of a range of technological options for the satisfaction
  of the identified, felt, needs.

  2. presentation of these options to the people

  3. enlisting the active participation of the people in the selection
  of that option or those options most appropriate to their development.

APPENDIX I

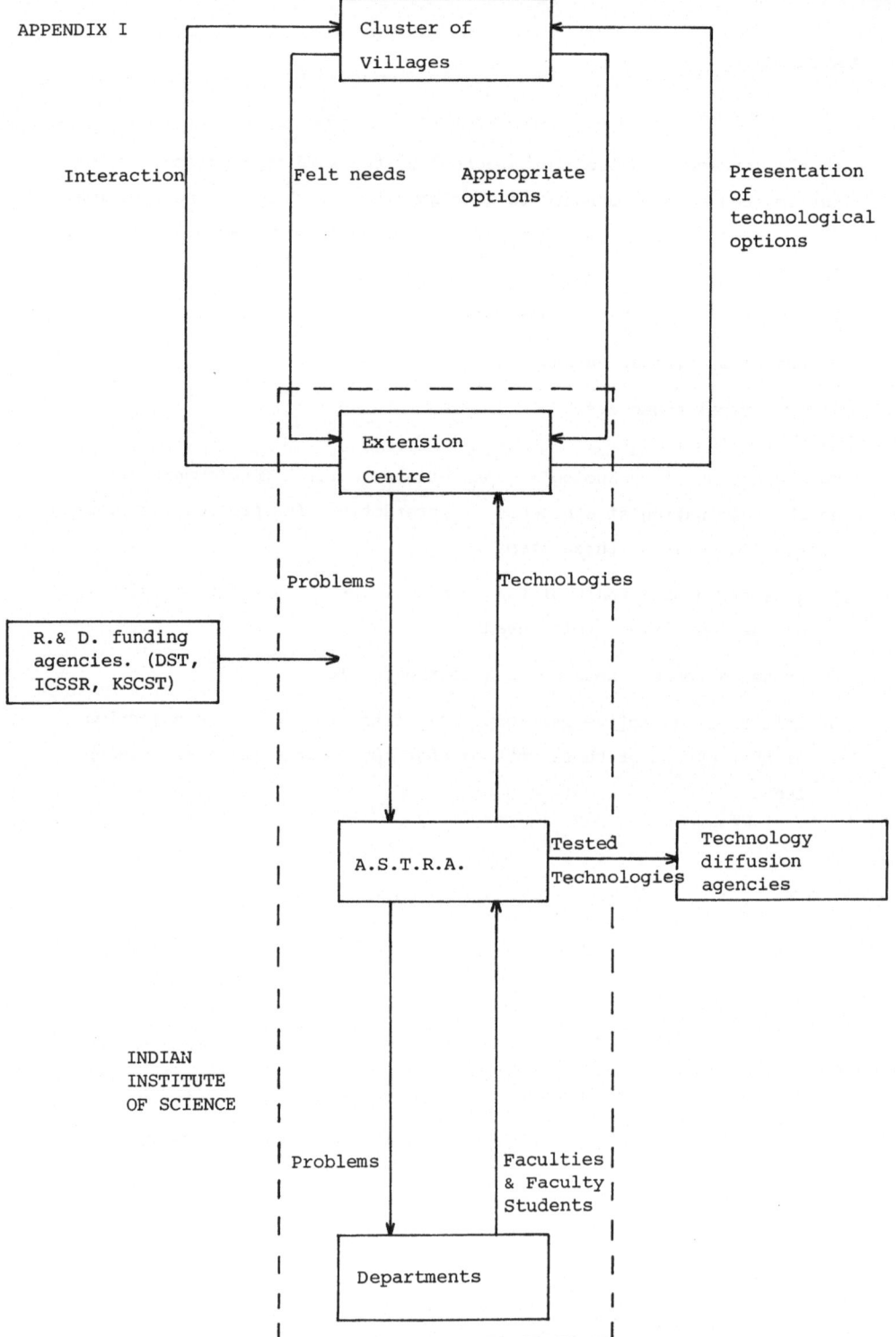

138

THE TECHNOLOGY-SOCIETY INTERACTION SCHEME FOR DEVELOPING COUNTRIES

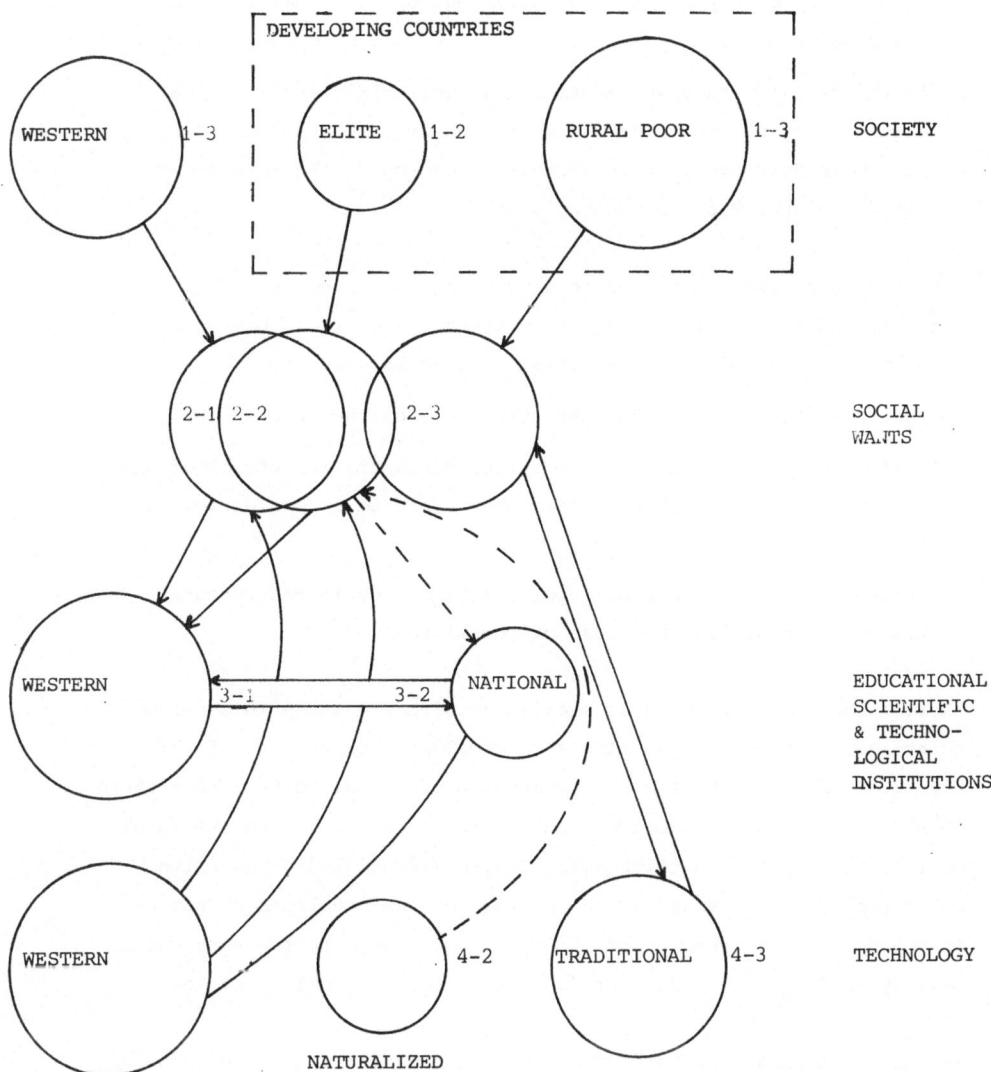

This appendix also refers to a paper presented by Amulya Kumar N. Reddy at the "Appropriate Technology Workshop" April 24 - 25, 1978 organized by the Karnataka State Council for Science and Technology for teachers from Engineering Colleges.

It is necessary to make the following comments about the schematic representation in the figure of Appendix II.

1. Little significance must be attached to the sizes of the circles,
   though

   a. in the case of the row: Society, the circles 1-1, 1-2, 1-3 have
      been drawn very approximately according to the relative sizes of
      the population; and

   b. in the case of the rows: Educational, Scientific and Technical In-
      stitutions, and Technology, the circles 3-1, 3-2, 4-1 and 4-2
      have been drawn very approximately according to the relative mag-
      nitudes of the R and D expenditures.

2. The problem of the urban poor in developing countries is indeed a
   serious, and more visible, problem. Nevertheless, the urban poor
   have not been included in the present scheme because

   a. their numbers are much smaller than those of the rural poor;

   b. being subject so much to the powerful demonstration effect of the
      life-styles of the urban elites, they share to a considerable ex-
      tend similar aspirations; and

   c. their survival in urban metropolises generates infrastructural
      requirements similar to those of the elites.

3. Whereas there is a tremendous overlap between the wants in developed
   countries and those of the elites in developing countries (cf. circles 2-1 and 2-2), it is a characteristic of a dual society that there
   is virtually no overlap between the wants of the elite and the rural
   poor in developing countries (cf. circles 2-2 and 2-3). The elite
   wants tend to be modelled on the pattern of the developed countries,
   in contrast to the rural poor who wants correspond to the very basic
   minimum needs of food, shelter, clothing, health, employment, etc.

4. In dual societies, the bulk of the decision-making is in the hands of
   the elites who are, therefore, responsible for the filtering process
   which selects out some wants for onward transmission as demands upon
   the educational, scientific and technological institutions and shelves
   other wants by giving them relatively little attention. In most cases,

this elitist filtering process functions in such a way that:

a. the wants of the elites are almost wholly transmitted as demands requiring technological answers, and

b. the wants of the rural poor are largely ignored even though they are an expression of urgent minimum needs.

Since it is the satisfaction of these basic needs which constitutes the essence of development, it follows that an elitist filtering process is incompatible with development.

5. The demands of the elite are picked up by educational, scientific and technological institutions through the agency of industry in the developing country and industry in developed countries, both of which sense in these demands a major market. It is important, however, to note that industry in developing countries is of two categories:

a. indigenous industry which derives its technology from the national educational, scientific and technological institutions; and

b. industry which may be owned by government, native entrepreneurs or by multinational corporations (or by two or three of these in different ratios), but which is based on imported technology generated in the institutions of developed countries.

Between these two categories, the linkage of the demands of the elite is very much stronger with the second category of local industry, viz., that based on imported western technology developed by the educational, scientific and technological institutions of the developed countries. This is why the strong linkage 2-2 — 3-1 is shown with a continuous line and the weak linkage 2-2 — 3-2 with a dashed line.

6. The operation of the filtering process to block the transmission of most of the wants of the rural poor, i.e., the basic minimum needs of the majority of the poverty-stricken population, from the educational, scientific and educational institutions is emphasised by the absence of a linkage between the circle 2-3 and either circle 3-1 or circle 3-2. Of course, the linkage is not zero- for instance, when the rural poor suffer from communicable epidemic diseases, the elites

are also vulnerable, and such needs of the rural poor are obviously responded to effectively. Thus, the filtering process is not conductive to rural development which in developing countries must constitue a major aspect of the development process.

7. In the absence of institutions to develop technologies to meet the needs of the rural poor, the latter have no choice except to fall back on traditional technologies based on the reservoir of empirical knowledge accumulated through the centuries (cf. the linkages 2-3 — 4-3 and 4-3 — 2-3).

8. There are very strong linkages 3-1 — 3-2 between the educational, scientific and technological institutions of developed countries and those in developing countries, the latter being modelled very closely on those of the former. In fact, these institutions in developing countries derive their emerging ideas for research and development, its trends and fashions, its stream of inspiration, its experimental techniques and instruments, its criteria of excellence and its source of recognition from the counterpart institutions in the developed countries. Superimposed on this process is the fact that developing countries receive education and training for a large proportion of their personnel, and in many countries a large influx of expatriates and "experts", from developed countries.

9. In the generation of technology, the educational, scientific and technological institutions of developing countries invariably start with imported western technology as a starting-point and as a model, hence the linkage 4-1 — 3-2. Thus, they emerge (linkage 3-2 — 4-2) after a process of imitation, adaption and innovation (the innovation, rarely!) with a technology which has been described as *naturalized*, i.e., adapted western technology.

10. The satisfaction of the demands of the elite is much more through western technology (this strong linkage is shown by a continuous arrow 4-1 — 2-2) than through naturalized technology (this weak line is shown by the dashed arrow 4-2 — 2-2).

Session 3: The Framework for Appropriate Technology.
The Role of Appropriate Technology in (Rural)
Development

APPROPRIATE TECHNOLOGY : THEORY, POLICY AND PRACTICE

*Marilyn Carr**

Introduction

This paper seeks to summarise some of the concepts surrounding appropriate
technology and to examine their validity in the light of existing socio-
economic and institutional constraints. It does not attempt to offer solutions
but, rather, to provoke discussion of some of the main issues affecting the
identification, adoption and diffusion of appropriate technologies. The paper
is divided into three sections. The first reviews the concepts of appropriate
technology. The second discusses the need to create the right climate to
enable the concepts to be translated into concrete results. The third section
discusses the role of Appropriate Technology Institutions in this process.

Review of Appropriate Technology Concepts

Despite the enormous interest expressed in appropriate technology and the growing acceptance of the concept by planners, both in aid-giving and aid-recei-
ving countries, there is still some confusion about what is meant by the term
and what is involved in the concepts.

Generally speaking, there is a certain amount of agreement that - in the
developing countries - the most appropriate technologies in the prevailing
circumstances are likely to be more productive than the often highly labourintensive but inefficient traditional technologies and less costly and more
manageable than the large-scale, labour-saving and capital intensive technologies of industrialized societies. These technologies will generally be
cheaper and smaller, giving a wider, more equitable distribution of capital
investment; they will create employment - providing work opportunities in

I.T.D.G. Ltd, United Kingdom

areas where people live; they will foster the use of local capital, skills and raw materials and reduce reliance on the importation of these factors; and they will produce goods primarily for local consumption. These are the range of technologies which are commonly referred to as 'intermediate' technologies. It would be wrong, strictly speaking to equate this expression with 'appropriate' technology; the latter being more wide ranging in its application, in that there are some circumstances in most countries where the appropriate technology may be large, expensive and labour-saving. However, in practice, the expression apprapriate technology has come to be generally used as being synonymous with intermediate technology, and it will be used in that sense in this paper. But the essential difference must be forgotten because it is but a short step from the argument that 'large' can sometimes be appropriate to the belief that appropriate is therefore 'large'.

Criticisms have been raised against these concepts of appropriate technology by some who interpret the expression as implying that developing countries should be satisfied with something inferior, second best and less efficient; and that two different standards are being established - thus creating a 'technological gap'. In return, proponents of appropriate technology have strongly pointed out that this is a misconception. For example, Hans Singer has argued that :

"The gap exists in the fact that some countries are poor while other countries are rich. The task is to reduce or eliminate this gap - the economic gap. Different technologies will serve to reduce the economic gap and hence, ultimately, to eliminate the need for different technologies, as just argued. If we misdefine the problem by declaring that the gap is a technological gap, and then try (disregarding the economic gap) to apply exactly the same technology to the two groups of countries, the real economic gap will widen further instead of narrowing. Thus, to say that the application of the concept of appropriate technology perpetuates the gaps between rich and poor countries is a travesty of the true position.'

The arguments put forward in favour of appropriate technologies are good ones, but tend to simplify what is, in reality, a very complex situation.

Hans Singer, 'Technologies for Basic Needs', Geneva, ILO,1977. p9.

146

For example, not nearly enough thought has been given to the vital question 'appropriate for whom?' Within a developing country, interests and needs can and do differ widely between people in urban and rural areas; large-scale and small-scale industrialists; large-scale and small-scale farmers; land-owners and landless labourers; communities and individual households and even members within a household. Thus, although it may be possible to talk of technologies which are most appropriate in terms of national development objectives - it must be realized that no technology can be appropriate for every group within the country.

Thus, the development objectives of the past which emphasized maximization of output and rapid industrialization involved technologies which are appropriate for the urban elite and the wealthier landowners, but widened the gap between them and the majority of the rural poor. Current development objectives are more concerned with reduction of unemployment and poverty and involve very different technologies. These are more likely to be technologies which are appropriate to the mass of small artisans and farmers in the sense that they have easy access to them with their limited cash resources and can use them to raise productivity levels in the rural and informal urban sectors.

To a large extent, the political and economic climate in developing countries has been, and usually still is, one which favours the adoption and use of those technologies which are appropriate for large industrialists and land-owners.

Within this framework, it is very feasible for large-scale entrepreneurs to adopt and use those technologies which are highly appropriate to their needs but which do little to alleviate the position of the poor. There is usually no incentive for these people to adopt those technologies which are normally referred to as being 'appropriate' - in fact, the incentive mechanism and the country's infrastructure usually make it *infeasible* to do so. Compounding the problem is the fact that the distribution of income is usually very un-equal and the effective demand is for products similar to those produced in industrialized countries which tend to have a capital-intensive technology. The products of what is termed 'appropriate' technology are more likely to be those which can be produced and used by the rural and urban poor, but while the distribution of income remains unequal, demand for these products remains non-effective.

147

In this context, it would seem that a change towards an 'appropriate' incentive/disincentive mechanism; an 'appropriate' infrastructure; and an 'appropriate' income distribution must accompany the adoption and widespread use of 'appropriate' technologies. Those who believe that the technologies appropriate for developing countries are those which are low-cost, capital-saving, employment-generating, decentralized and sparing in their use of imported goods and skills should perhaps pay more attention to what is involved in creating the economic climate which is necessary if these technologies are to be adopted and used on a widespread basis.

Creating the Right Climate

The identification of technologies which are appropriate to conditions prevailing in the developing countries will be of little use unless they are applied on a widespread basis. Also, there is little point in simply advocating the increased use of these technologies without offering suggestions as to how this can be accomplished.

What might appear to be desirable technologies can be introduced and proved to be technically and economically viable, but their diffusion throughout the country will be severely limited unless the socio-economic climate is such that it will encourage both the manufacture and the use of the new technologies.

In theory, it should be possible to devise a comprehensive package of incentives and disincentives to fit any circumstances. Thus, if a government is concerned with reducing unemployment and poverty, it should be possible to devise a package of policies which lead to the adoption of technologies which are appropriate in the context of achieving these objectives. In this particular case, the policies involved will be ones which induce existing producers (e.g. large-scale manufacturers and landowners) to adopt - on economic grounds - those technologies which benefit the poor and unemployed. They may also be policies which increase the purchasing power of the rural and urban poor so that an effective demand is created for goods which lend themselves more readily to production by low income producers who are more likely to use 'appropriate' technologies. In practice, the extent of government intervention depends not so much on what government might wish to achieve, but how far it can go without losing the support of such powerful groups as land-

owners, large industrialists and the urban elite. As far as government is concerned, the extent to which it can achieve its stated development objectives
(e.g. reduction of unemployment and poverty) without losing the support of
those who - in the short term - control the economy, will determine what is
and what is not appropriate.

Government policy with respect to import licensing, industrial licensing, price
control and many other issues can and often does stop the spread of profitable
small-scale technologies. One example of this is the development in Ghana of
an intermediate technology for producing animal feed from brewer's spent grain.
This technology, which consists of simple hand-pressing and solar drying, produces animal feed at a much lower cost and with less use of scarce foreign
exchange than the advanced technology normally used by the breweries. However,
the decision of the government to allow the breweries to import sophisticated
drying equipment looks as if it may:
a. put existing small-scale entrepreneurs using the intermediate
   technology out of business (by eliminating the supply of spent
   grain);
b. eliminate the possibility of other small entrepreneurs joining
   the market (for the same reason); and
c. lead to a rise in the cost of animal feed.*

Since the intermediate technology results in lower unit costs and more employment than the advanced technology, and also provides a source of income for
small entrepreneurs, it would have been more appropriate from the point of
view of the consumers of animal feed, the unemployed and the small entrepreneurs to encourage this by refusing to grant an import license to the breweries. It would also have been desirable from the government's point of view
in saving foreign exchange. This, however, would not, in the existing circumstances, have been the most appropriate policy from the point of view of the
breweries. The questions to ask here are :
a. could indirect government policies such as price controls alter circumstances so that the breweries no longer find it desirable to use the
   imported technology, or would direct measures such as import or indus-

*Sally Holterman, 'Intermediate Technology in Ghana: The experience of Kumasi
 University's Technology Consultancy Centre'. Intermediate Technology Publications Ltd. August 1979

trial licensing be necessary; and

b. to what extent would such policy measures be feasible from the
government's point of view?

Those who argue for the need to create the right social-economic environment
before desired changes in the lot of poor people can take place, often do so
in a negative fashion, arguing that nothing can be done unless 'the system'
is changed. That argument is not the basis upon which the above considerations
are put forward. It is merely that in planning the introduction and diffusion
of appropriate technologies, there is a need for governments to look at the
different interventions they can make and to examine their feasibility as well
as their possible value and effect in achieving desired results; by the same
token, outside agencies advocating the use of appropriate technologies must be
aware of the need for appropriate interventions and must be able to assess the
extent to which such interventions are possible in the circumstances.

The Role of AT Institutions

Institutions, including those dealing with appropriate technology, are part
of the infrastructure package in development plans and the package is, itself,
one instrument to be used in creating the right climate. Institution building
must be seen within the context of the whole infrastructural framework. For
example, how do we balance the need to create specific AT institutions against
the need to instill AT concepts and capability into the institutional fabric
as a whole?

Although in the long term it may be ideal to have the concepts of appropriate
technology incorporated into the institutional framework as a whole, there is
probably a need for specific AT institutions in the short term until the concepts become part of the conventional wisdom. If these specialist institutions
are successful, there should eventually be no need for their existence.
But, in creating them, to what extent are we perpetuating the concept that
appropriate technology is something different from technologies which are appropriate?

At present, AT institutions exist at the local/national, regional and international levels. There are points for and against creating institutions at any
of these levels. One very good reason for the development of local AT insti-

tutions is that conditions in each country (or, in some cases, region of a country) are not only unique but also constantly changing. Generally speaking, therefore, only an institution actually located in the country can fully understand the nature of the problems to be solved, keep up-to-date with changing economic conditions and design or adapt technologies accordingly.

One very good example of this is the case of the involvement of the Technology Consultancy Centre (TCC) in soap making in Ghana. Initially the TCC designed an intermediate level soap plant which was technically and economically viable and could be operated by small entrepreneurs in an attempt to meet existing soap shortages. When the supply of imported caustic soda (an important ingredient in soap making) became scarce, the TCC was able to respond by developing small-scale plants for producing this substance from the by-product of a local factory. Finally, with the supply of palm oil and perfume now becoming irregular and scarce, experiments are going on with a view to finding alternative sources for the raw material.*

Another good argument in favour of local or national institutions is that they help to build up the local technological capability and expertise which is needed if reliance on imported technologies is to be reduced. Although it is sometimes argued that duplication of effort occurs when numerous local institutions carry out similar research and development work, perhaps this should be looked at in the more favourable light of being a part of the process of building up the technological capability in each country. Local and national institutions can also respond more easily to needs such as training, maintenance and repair services and credit facilities which accompany technological innovation. For instance, in the case of the TCC's involvement in soap making, necessary back-up has been provided for training plant operators, assisting entrepreneurs to obtain bank loans and helping them through initial operational difficulties.

Local or national institutions can also respond quickly and efficiently to technical enquiries. Very few small entrepreneurs or field workers have the necessary skills to write a technical enquiry giving enough details to enable an information centre to respond in a meaningful way. If the information centre is located in the same country, then further details can be requested and

*Sally Holterman, op. cit.

acquired more quickly than is the case with a regional or international centre. A local institution is also obviously better able to assist with implementation of solutions.

Local or national institutions do, of course, have their drawbacks and limitations. Precisely because they adopt an in-depth, practical approach, their impact - although a direct one - is limited to their own fairly small 'catchment' area. Given the limited amount of manpower and funds available for AT work, there are, therefore, large areas within countries and often entire countries which do not have access to the advice and services of a local AT institution. This can be interpreted as an argument for having regional/international institutions which have centralized information systems and give at least some (although obviously much less thorough) assistance to a greater number of people.

Another proposed role for regional/international institutions is that acting as a central data bank supplying all local institutions with information about each other's activities. The question which needs to be asked is whether this is more or less efficient than providing the necessary support to allow the staff of the local institutions to visit each other and create their own information network.* When considering this, it would be useful to evaluate the extent to which existing regional centres have related effectively with small national centres and groups within their own region, and the extent to which links forged directly between local centres have been effective.

Although it is generally true to say that local/national centres can achieve many positive results at grass-roots level even without the help of regional/international centres, the same regional/international centres are able to achieve very little in positive terms unless local/national centres exist to act as intermediaries by interpreting local needs and constraints and introducing the required imporvements. Given this, perhaps more thought should be given to the need for initiating and supporting local/national centres rather than setting up even more extra-territorial centres for the transfer of technology.

---

*For example, external assistance has enabled staff from AT Institutions in The Gambia and Upper Volta to visit the TCC in Ghana.

STATEMENTS

- An appropriate technology should be described in terms of appropriate for
  whom, where and when? There is also a need to give more consideration to
  some of the less widely used criteria of appropriateness such as systems
  independence, the image of modernity and the evolutionary capacity of a
  technology.

- If appropriate technologies have already been designed and these are not
  used on a widespread basis, this may be due to the fact that the correct
  economic framework is absent.

- The role of national appropriate technology institutions in developing
  countries can relate to technology policy in two important ways:

  1. They can react fairly quickly to changes in technology
     policy and develop or adapt technologies accordingly;

  2. They may be able to influence national policies in ways
     which encourage a more widespread use of appropriate
     technology.

TECHNICAL CHANGE, EMPLOYMENT AND DISTRIBUTION IN LDC'S

Staffan Jacobsson *

TRENDS IN MANUFACTURING OUTPUT AND EMPLOYMENT IN LDC'S AND DC'S

When discussing unemployment problems in Less Developed Countries (henceforth LDC's), it is absolutely essential to start off by assessing the *magnitude* of the problem. We shall first concentrate on the manufacturing sector as the traditional view of the evolution of the structure of employment in economic development has been that of the manufacturing sector gradually absorbing rural labour and eventually transforming the pattern of employment into something resembling that of today's Developed Countries (henceforth DC's).

As has been noted by several authors[1], manufacturing output has grown much faster than manufacturing employment also in LDC's. For data on five selected countries see Table I[2].

Table I : Changes in output and employment in per cent per annum

	Output	Employment
Philippines	5.4	2.0
India	5.4	2.6
Korea (Rep. of)	21.1	12.7
Kenya	8.2	6.5
Peru	6.9	4.1
Brazil	11.3	4.9 (Only Sao Paolo area)

However, in isolation, the magnitude of the difference between these figures is not so fruitful to focus upon. Instead, it is more revealing to compare the yearly growth in demand for labour with both the size of the manufacturing

* *University of Lund, Sweden*

sector in relation to other sectors, and with the rate of labourforce increase. In Table II[3] we have assembled some data which give a rather interesting perspective on the employment problem. It shows that on the basis of past trends, not even the yearly addition to the labour force has been absorbed by the manufacturing sector in *any* of the countries. Indeed, apart from Korea, the supply of jobs through the expansion of the manufacturing sector was *extremely* insufficient in relation to the additional jobs needed stemming from the growth of the labour force, not to speak of the vast number of already unemployed. This is the first fact we have to keep in mind when discussing unemployment problem in LDC's.

Table II

	$\frac{Em}{Em}$	$\frac{Em}{Et}$	$\frac{L}{L}$	$\frac{Em}{Em}$ needed to absorb $\frac{L}{L}$		
	(1)	(2)	(3)	(4)		
Philippines	2.0	11.4	2.8	24.5		
India	2.6	9.5	2.1	22.1		
Korea (Rep.of)	12.7	13.2	1.8	13.6		
Peru	4.1	13.2	2.9	21.9		
Brazil	4.9*	17.8	2.9	16.2		
Kenya	6.5	16.3**	3.5	21.4		

$\frac{Em}{Em}$ is the yearly increase in manufacturing employment

$\frac{Em}{Et}$ is manufacturing labour force of % of total labour force

$\frac{L}{L}$ is the labour force increase

Furthermore, when we are dealing with unemployment problems, we will also have to take into account the trends in technical change in the industrialized world.

* *Only Sao Paolo area,*   ** *wage employment exclusing agriculture*

This is important as the overwhelming majority of the world's technology is
produced there and there is nothing that says that the technological dependence of LDC's on DC's will radically diminish during the next decades. In
Table III[4] we have reproduced the data on trends in manufacturing output and
employment in the 'EEC Five' countries. As one would expect the LDC's to experience a lag in the vintage of their technologies, it seems fruitful to include a fair number of years in the analysis.

Table III : Annual average rates of change in percent per annum

	55-60	60-64	64-69	69-73
Employment	2.58	1.66	0.53	0.72
Output	6.85	6.55	6.51	5.39

Similarly, in Sweden the trends were as follows[5] :

Table IV

	50-55	55-60	60-65	65-70	70-75
Output	2.5	4.8	6.9	5.1	2.4
Number of workhours	0	-0.2	0	-1.8	-1.8

The note in the Swedish case, the number of employed have been substituted
for by the number of hours worked. As there has been a substantial reduction
in the workday during the last decades, the figures in Table III underestimate
the real trends in the reduction in labour input. This is important for
LDC's as it is not unreasonable to assume that the working day is faster reduced in DC's than in LDC's due to different relative strength of capital and
labour.

Thus, in the post-war period, and in particular since the early 1960's, there
has been a systematic downward trend in employment generation from a given
rate of increase in output. Both Freeman[6] and Clark[7] argue that this trend
can be, to a great extent, explained by an increase in the relative share of
rationalization investment relative to expansionary and replacement investment. Clark, for example, writes after examining the trend in annual change

in British manufacturing employment per unit of investment that[8] :

"The implication then is that there has been a consistent decline in the employment generated by a unit of investment since 1950 and that since about 1965, the effect has been predominantly a labour displacing one."

Whilst part of the change in labour input versus output can be explained by a structural shift of relatively labourintensive processes to LDC's, the magnitude of the changes suggest that it is most likely more closely related to the direction of technical change[9]. The figures given in Tables III and IV would then reflect an intensified process of laboursaving technical change in DC's, a 'jobless growth'. Consequently, if we are right with the assumption of a significant technological lag between LDC's and DC's, this direction of technical change will later be reflected in LDC's, making the possibilities to absorb the yearly addition to the labourforce through expanding the manufacturing output even smaller. This is the second fact we have to consider when discussing unemployment in LDC's.

THE NATURE OF TECHNICAL CHANGE AND THE SCOPE FOR LABOURINTENSIVE MANUFACTURING TECHNOLOGIES

Lately, it has become very common to link the very large and growing[10] unemployment problems in LDC's with the inadequacy of the manufacturing sector to absorb labour. The people and organisations who make this link, e.g. ILO and UNIDO, note that the distribution of the non-socialist world's research and development resources (henceforth R&D resources), is very heavily skewed in favour of the DC's. Then they argue that, in the words of Frances Stewart[11],

".... the new technique element has been almost all concentrated on the capital-intensive end of the spectrum because of the concentration of research and development, and manufacture of new techniques, in the capital-abundant developed countries".

Thus, they argue that as the technology emanating from DC's is designed in the light of the investment resources they can afford, practically the whole world technology reflects the conditions prevailing in the developed capital abundant economies.

The proponents of this view then take the argument one step further and argue that labour intensive technologies would be both feasible and efficient if only sufficient R&D resources were allocated for the purpose of developing them. That is, much of the employment problem in LDC's could be solved by re-allocating the world's R&D resources and developing labour intensive alternatives in the manufacturing sector. This is much what the present debate is about in international organisations and among development economists. The policy implications on this view will then go along the lines of "how do we get appropriate (labourintensive technologies generated and disseminated?" This view of both the cause and solution to unemployment problems is also shared by the Indian National Paper for UNCSTED [12]:

"Thus, the technologies of the developed countries embody the characteristics of satisfying ..... capital-rich, labourscarce situations ... This feature of the technologies has been largely responsible for inhibiting the generation of employment despite massive investment on industrialization ... if only capitalintensive technologies are chosen then the objective of employment generation cannot be achieved. This ... dilemma can only be resolved by adapting a balanced mix of capitalintensive and labourintensive technologies ..."

We would argue that this way of looking at the unemployment problem is basically wrong; that is, it is not relevant to discuss unemployment in terms of choice of technique in the manufacturing sector.

As opposed to the previous view as expressed by F. Stewart, where the degree of capitalintensity had no relation with the rate of technical progress, we would argue that technical progress is necessarily *more strongly* associated with a high capitalintensity, a high capital/labour ratio.

Whilst the basis for the proposal that labour intensive technologies can be developed on a large scale in the economist's conceptualization of alternative techniques in terms of different *quantities* of capital and labour, we would suggest that there are extremely important *qualitive*[13] differences between the two factors of production which make the development of labourintensive technologies less feasible.

To our knowledge, the first economist or social scientist, who first pointed out the qualitative differences between capital and labour was Marx[14]. The distinctive feature of what he called large scale modern industry, was that the characteristics of the worker and his *physical limitations* did not constitute a limiting factor in the design of the production process. From studying the history of technical change, one may as Marx did,draw the conclusion that technical change to a large extent is a process of *overcoming*, through increasing the capital intensity of the production process, the restrictions set by the *properties of human labour*. In line with his analysis, we would argue that the physical properties of labour are quite different from those of a machine. In relation to a machine, a person is variable - implying uneven quality in performance, a person lacks strength - with obvious implications, he cannot achieve the same precision as a machine which is absolutely fundamental for any machinemaking activity, he cannot stand extreme heat - heat is used in production of various important processes such as steel production and in chemicals, he is slow - which implies that in *any* industry which produces above a certain minimum level output a machine will be superior to a person(s). The development of labourintensive technologies will necessarily mean that the production process will, to a greater extent, be based on *direct* labour and, consequently, that it will be limited by the properties of human labour described above.

As a consequence, we would argue that the opportunities for developing labour intensive, efficient techniques in various industries depend on the incidence of 'technical barriers'[15], that is, functions which due to human limitations either cannot be performed manually or can be performed only in a very inefficient way by direct labour. One may distinguish between absolute and relative barriers. An absolute barrier exist, for example, in steel making where extreme heat is a prerequisite for production. In industries having absolute barriers, capital intensive production methods are the only feasible ones. An example of a relative technical barrier is human speed in computing which, even if a feasible alternative, is far inferior to the use of microprocessors. In this case a relatively capital intensive technique will due to the vast increase in productivity resulting from overcoming the 'barrier' be superior to any alternative. The consequent increase in productivity due to overcoming a barrier will be *localized* in the capitalintensive end of the spectrum since the increase in productivity occurs only as a result of the use of a technique which *requires* a high capitalintensity.

The important point to realize is that this process of overcoming barriers and disseminating the technical solutions to other industries has been going on ever since the Industrial Revolution and that it clearly continues today. In the 19th century, the demands for *precision* meant that the development of machinetools such as the grinding *machine*. Today the developments in the electronics industry and the increasingly wider application of microprocessors in various industries, including also typical batch production industries such as mechanical industries, is perhaps the prime example. As the diffusion of microelectronics in industry and services will have very important implications for the issues dealt with in this essay, we will elaborate on it.

Any system which involves the processing of data, decision making or control of systems or equipment is a candidate for the application of microelectronics[16]. The key to the whole issue is the *extreme superiority* of the microprocessors over humans in terms of reliability (variability), speed and flexibility in information processing. While the most apparent use of microelectronics will be in precisely the information industry - such as telecommunications and offices - they can also be used for controlling individual pieces of equipment in manufacturing as well as controlling subsystems and whole production systems. It has been suggested that any task involving logic can be undertaken by a suitably programmed microprocessor linked to a suitable device. A list (not exhaustive) of these tasks include[17]:

1 controlled movement of materials, components, products
2 control of process variables
3 shaping, cutting, mixing, moulding etc. of materials
4 assembly of components into sub-assemblies and finished products
5 control of quality at all stages of manufacture by inspection. testing or analysis
6 organisation of the manufacturing process, including design, stock-keeping dispatch, machine maintenance, invoicing and the allocation of tasks.

Thus, the application of microelectronics may be found in sectors with mass-production techniques as well as in sectors with batch production.Furthermore, as point (6) indicates, the existence of cheap, fast and reliable information processing systems opens up the possibility of the eventual realization and implementation of the complete integrated automatic factory. Indeed in Japan, prototype unmanned factories have already been designed.[18]

The effects on employment will be particularly strong in the information sector where, so far, relatively little capital has been invested in improving labourproductivity. However, also labourintensive parts of the actual production in manufacturing are likely to be transformed into more automated units. As the laboursaving consequences of the introduction of microprocessors have been the target of most attention, we will not elaborate on it further than noting that the trends in industrial employment and output showed in section 1 will clearly continue in the same direction and very possible in an intensified way.

What is not so generally noted is the *capitalsaving* potential of technical change involving microelectronics. We will exemplify with the machinetool industry. In metalworking industries, batchproduction dominates over flow-line techniques with an associated low efficiency through poor machineutilization. Numerically Controlled machinetools, NC machines, constituted a first attempt to increase the efficiency in this sector. With these machines, the control signals containing the information needed to produce the part are fed into the machine as the operation is performed. The control signals imitate the instructions given by a skilled machine operator, but with much greater speed and precision. By changing the control tape, an NC machine can be quickly switched to the next job which may involve a totally different sequence of operations. In this way, the downtime - the setting time - of the machinetool is reduced which is very important for machineutilization in small batch production work. By replacing the still relatively inflexible hardware circuitry in the NC machines by software mini or micro computers - i.e. producing Computerized Numerically Controlled machine tools (CNC) - the versatility and flexibility of the machinetools are considerably enhanced.

The capitalsaving nature of technical change in this sector stems not only from increased machineutilization. CNC and DNC (Direct Numerical Control which involves one computer controlling several machinetools) also increase quality, for example in precision lathing. They also increase the throughput and reduce inventories which saves capital embodied in materials. Furthermore, they allow for in-process quality control which make possible an early localization of mistakes and correction of the process variables through the electronic feedback system. This latter source of capitalsaving is of considerable importance for also process flow techniques, for example in paper pulp and glass production, where work in progress often constitutes a very important part of total capitalcost.

The automation consequent upon the overcoming of the human limitations of speed, variability and flexibility in information processing is therefore not only labour saving but also capital saving. In that sense it is appropriate not only for DC's but also for LDC's. The very important point to notice is that this kind of overall factorsaving technical change would not have been feasible to undertake with high labourintensity in information processing and control. Equivalently, it is most likely so that the reaping of the advantages of microprocessors is necessarily associated with a higher degree of automation.

Thus, we would argue that the development of capitalintensive techniques has not, and will not primarily be the result of the present skew distribution of the world's R&D resources. Instead, technical change should to a greater extent be seen as a process of overcoming human limitations to production through increasing the capital intensity of the methods of production. Consequently, labourintensive alternatives would, even if more R&D resources were allocated to them, become increasingly inefficient.

Now, whilst the development of capitalintensive technologies may not primarily be the consequence of a skew distribution of R&D resources, it is quite plausible that there are different pressures working simultaneously for developing such technologies.

Neoclassical economic theory teaches us that the direction of technical change is determined by the relative factordistribution in the economy in question. This view of the cause of increased capitalintensity could of course be partly valid in that a process of substitution of capital for labour takes place simultaneously with improvements in the technology wich can only be incorporated into machinery. To the extent that this is the case, a technological strategy based upon the selective mechanisation of earlier techniques would seem appropriate for LDC's.

Korea seems to have successfully followed such a technological strategy in textile machinery and adapted a semiautomatic technology, e.g. by increasing the size of the shuttle, so that it is far superior to imported automatic looms at the relative price structure of capital and labour in Korea.[19] The

162

employment generation of the use of this indigeneous technology would still be more than ten times higher than the employment generation associated with the use of the imported automatic machine.[20]

Clearly, the upgrading of an older technology, using mechanical as contrasted to electronic, knowledge has been a very profitable strategy to follow for Korea.

What are then the inplications of this Korean experience? How far can it be generalized and what are the restrictions to this strategy?

Firstly, it is limited in its applicability to the degree that increases in the capitalintensity reflects a substitution process and not a process of overcoming human limitations.

Secondly, the incorporation of electronics and other advanced techniques on a wider scale may change the relative competitiveness of the adapted technologies. Thus, improvements based on mechanical knowledge may, in the long run, become inferior to improvements based on higher level knowledge.

Thirdly, there are very important internal and external *non-technical* restrictions for the implementation of an adaptive strategy. To begin with the *internal* ones, we may first note that contrary to the case of Korean textiles, the *quality* and the *type* of the product often changes when more labourintensive techniques are used. This change in the characteristics of the product will for two reasons mean that labourintensive techniques may not be developed/chosen;

a it is now established that consumer preferences for branded, high quality consumer goods are very strong in LDC's,

b it is also clear that local firms which want to compete with multinational companies (henceforth MNC's) on the home market often have to adopt the technology associated with the production of very specified products in order to survive. This is associated with the fact that competition through product differentiation and marketing is more frequent than price competition.[21]

A second internal restriction results from the fact that increased mechanisation is a way for the capitalist to *control* the production process more

fully. The desire/necessity is as old as capitalism itself. It is enlighte-
ning to quote one of the classics in the history of technology; Ure[22], who
describes the consequences of a strike in the cotton industry in the following way :

*"...During a disastrous turmoil of the kind at Hyde, Stanely Bridge, and
the adjoining factory townships, several of the capitalists, afraid of their
business being driven to France, Belgium and the United States, had resource
to the celebrated machinists Messrs. Sharp and Co. of Manchester, requesting
them to direct the inventive talents of their partner Mr. Roberts, to the
construction of a self-acting mule."*

When Roberts had successfully completed his task, Ure concludes:

*"...This invention confirms the great doctrine ... that when capital enlists
science in her service, the refractory hand of labour will always be taught
docility."*

Thus, due to the capitalist's constant fear of strikes and labour troubles,
he has an objective interest in developing more mechanised production methods.
Mechanised methods of production generally mean that a deskilling process
takes place, i.e. the workers will not be able to control the production process even if they so desired and of course, it also reduces the number of
workers involved.

Consequently, we may conclude that there are important obstacles on the
demand side for the use of more labourintensive technologies, and thus the
implementation of an adaptive strategy.

As to the *external* restrictions, the trade strategy of the country involved
will be crucial. An exported industrialization process, i.e. where the country produces for the markets in DC's, will impose similar contraints in terms
of quality on the choice of technique and technological strategy as they did
in the internal case.

A further very important external restriction lies in the large role multinational companies play in investment activities in LDC's. This dominance is
indeed very restrictive for several reasons. Firstly, as has been noted in the

debate, MNC's generally use the same technology when they direct invest as they do in their home economy. Such behaviour may be explained in several ways:

a the cost of developing new, more labourintensive techniques may be estimated to be higher than the potential gains from using such a technology,
   i.e. from utilizing the abundant and cheap labour in most LDC's.
b MNC's probably face different prices than investing domestic firms. In
   particular, they have access to cheaper capital which implies that there
   is less incentive for MNC's to develop labour intensive production methods
   even when such are feasible.
c A main advantage of MNC's lies in productdifferentation, access to brand-
   names and marketing. Consequently, also their choice of technique is restricted by the link between technique and product.

As a consequence of the failure of MNC's to develop labourintensive techniques, it is often argued that the development of such technologies must ori-
ginate in LDC's. However, we would argue that this view stems from a preoccupation with MNC's as technology producers.

As distinguished from MNC's, pure machine producers always have to produce
machinery according to the specifications of the customer. They would therefore be expected to develop machinery for also LDC markets. Indeed, the development of the Sensomatic loom may be a case in point.[23] A reorientation to
the acquisition of the technology from mainly machinery producers would however require that:

a such technologies are demanded, i.e. that the internal restrictions are
   overcome.
b that the MNC's investment activities are either restricted or controlled
   to ensure that they use more labourintensive technologies. The political
   dimensions of such a policy are of course enormous.

The second negative influence of MNC's can only be understood if we give up
the requirement that the labourintensive technologies have to be efficient at
world market prices. In this context it is absolutely crucial to understand
that in economies with severe balance of payment problems and a low output
capacity of the domestic capital goods sector - most LDC's today - there is
an objective upper limit to investment activities as any investment implies
a need for a machine of some kind. In other words, even if a LDC desired to

invest a lot, it would neither have the foreign exchange to buy foreign machinery nor the domestic capability to produce machinery. Consequently, in such a situation, *any* mobilization of domestic resources for machinery production will have a growth inducing effect. The use of labourintensive technologies, where they are feasible, could be part of a technological strategy aiming to overcome such a restriction on investment activity. This strategy would include both such inefficient labourintensive technologies and highly efficient imported or self-produced ones. However, such a strategy is incompatible with the dominance of MNC's in investment activities as they will outcompete any inefficient domestic production. At the same time there is no reason to assume that the basic problem of a weak balance of payments and low output capacity of the domestic capital goods sector will be overcome by allowing MNC's to establish themselves in the economy.[24] Presumable, this type of situation applies only to sectors of the economy where the quality of the output is variable, i.e. to the consumer goods sector mainly.

The general problem which has been treated in the last pages is the *institutional* contraints to the development and implementation of 'appropriate' technologies where these are technically feasible. That is, there must always be a person/organisation - a social carrier of technology[25] - which has both an interest and capability, both economically and politically, to implement more labourintensive technologies. Whilst this constraint is indeed very important in the manufacturing sector, its implications for employment is most likely fare more important in the rural sector. The next section will therefore focus mainly on these constraints in that sector.

In summary, we have argued that the proposition that the lack of development of labourintensive technologies in the manufacturing sector is the result of the skew distribution of the nonsocialist world's R&D resources is basically wrong. We have instead suggested that technical change may *mainly* be seen as a way of overcoming the restrictions set by the properties of human labour. These restrictions are particularily important when high quality of the product is important. They also imply that the output level is an important determinant of capitalintensity as machines are able to operate at much higher speed than humans. Whilst in some industries, labourintensive technologies are still feasible and efficient, we may see that such technologies may become inefficient due to two sorts of changes :

a: a market change, i.e. a change in output levels and/or specifications of
   the product,

b: process changes e.g. technical change which, as in the case of the development of CNC machinetools, may mean that low volume levels may not be a
   hindrance to capital intensive techniques any more.

We have further argued that there are very important institutional restrictions to the use of labourintensive technologies, stemming primarly from the
integration of LDC's with the world economy.

From the figures in the first section, we can calculate that if only the
yearly addition to the labourforce were to be absorbed by the manufacturing
sector, the labourintensity of new investment projects would on average have
to increase by a factor of 12.25 in the Philippines, 8.5 in India, 5.3 in
Peru and 3.3 in Brazil and Kenya.Therefore given the magnitude of the employment problem, the overriding implication is that 'appropriate technologies' may only in a marginal way reduce the unemployment problem. Instead,
we have to look for other ways of employing the fastly growing labour force
in LDC's. The most relevant question for employment generation will therefore
not be "how to generate and disseminate appropriate technologies", but "under
which conditions will the labour be absorbed in other sectors of the economy,
e.g. in agriculture?" As this is the sector where the majority of the Third
World's population live and work today, it seems natural to focus out attention first on that sector.

RURAL SOCIAL ORGANISATION AND EMPLOYMENT

The agricultural sector is generally looked upon as the sector which is
most flexible when it comes to variations in the capital/labour ratio. There
are different reasons for this. Firstly, whilst in industry a great deal of
the technologies are process flow technologies (which are particularly well
suited for high capitalintensity) this can obviously not apply to agriculture.
Secondly, the human limitations of precision/quality/strength etc. are not as
critical in agriculture as in industry. The only human limitation which may
apply is speed which may be an hindrance when introducing two or three crops
annually.

It seems widely accepted that migration from the rural to the urban sector
accounts for a great deal of the urban population growth, and therefore, of

the urban unemployment. It has been suggested that the wage differential and the subjective estimation of the probability to get a job in the urban sector will determine the rate of migration.[26] It has also been argued that :

"... in many developing countries ... the availibility of suitable land for agriculture is already very restricted and the median size of the farm small. In such cases, any acceleration in the growth of the rural labourforce can be expected sharply to increase pressures to migrate, further intensifying urban employment problems."[27]

What these suggestions have in common is that, in spite of some validity, they do not focus *directly* on the basic cause of rural poverty and hence migration : the extremely unequal access to resources, mainly land.
In *combination* with a high rate of growth of the population, the skew distribution of land creates a perfect breeding ground for an urban lumpen proletariat. We would therefore argue that the solution to the urban unemployment problem has to be found in the rural and not in the urban sector.

F. Stewart agrees on the latter proposition but seems to argue that appropriate technologies would solve the problem of urban unemployment by reducing the incentive to migrate to the urban sector. The use of appropriate technologies would imply a reduction in labour productivity in the urban sector and an increase in the productivity of labour in the rural sector. As a result, she argues that : [28]

"... the gap between the sectors is reduced and the consequence is that open unemployment may be reduced since there is a reduced incentive to switch between the sectors, with the greater productivity in the L-sector (rural sector)."

However, it is absolutely essential to understand that the productivity increases Stewart talks about have to be made inside a social organisation which combines the objective of productivity increases with the objective of distribution. These two objectives are not always fullfilled simultaneously as we will exemplify with the experience from the Green Revolution in India.

The High Yielding Varieties (henceforth HYV's) were introduced into a social structure with a very skewed distribution of land and political/economic power. For many reasons, the larger farmers also had preferential access to resources other than land.That is, the access to inputs such as fertilizers, seeds and capital and the realization of their benefits was confined by in- stitutional factors to only relatively large landowners.

As a consequence of the unequal access to resources, a cumulative movement towards unequal productivity increases was initiated and a process of capital- accumulation started to occur. The value of land increased substanitally which in turn increased the rents .and the desire of the landowners to cultivate the land themselves. As rents increase, poorer farmers were forced away from the land and the rate of eviction of peasants (e.g. sharecroppers) increased. In combination with the mechanisation associated with the spread of large scale profit maximising farming, these evictions had the effect of increasing the relative and absolute poverty in Punjab in spite of a substantial increase in production.[29]

Whilst the kind of rural organisation in Punjab may explain the effects of introducing a new technology on the distribution of employment and income, one also has to grasp the relationship between the rural and urban elites in order to fully understand the forces behind the type of rural change in a country such as India.

Available evidence strongly indicate that the Green Revolution has been asso- ciated with a socially inefficient allocation of resources.[30] In particular, there has been an extensive mechanisation on larger farms in spite of the existence of masses of underemployed labour. The main reasons for mechanisa- tion, all of which have nothing to do with social profitability, can be sum- marized as follows :

a   subsidized price of mechanised equipment,[31]

b   fear of labour troubles and dependence on casual labour at critical
    moments in production,[32]

c   the managerial and technical difficulties of employing a large labour
    force.[33]

The steering of resources to larger farms and the consequent deprivation of smaller farms of resources, should though be seen as a way for the urban eli- tes to get a sufficient *marketable surplus* of agricultural produce into the cities. As Dasgupta explains :[34]

"But of course, industrialists are deeply interested in a policy which will

ensure cheap food for their workers. The new agricultural strategy can be
seen precisely as resolving the essential conflict between industrial and
agrarian elites on this issue; cheap and increased food production is achie-
ved without the existing balance of forces in the countryside being upset."

Thus, the existing power structure of urban and rural elites had in common
the interest in strengthening the capacity of the larger farms. The criterion
for efficiency was thus changed to the quantity of marketable surplus that
could be made available which indeed shows how misleading it is to talk about
social profitability in a class society.

The general, very important, lesson from this Indian case is that there are
strong structural limitations on the choice of technology in the rural sector.
These limitations are not confined to the kind of rural social organisation,
but include also the relation between the rural and urban sectors. These
limitations have,to be identified *first* in any analysis of the real possibilities for labourabsorption in this sector.

We would thus argue that rural 'development' in the framework of an emerging
capitalist mode of production in the rural sector, where the elites have
strong common interest with the urban elites, will most likely be associated
with an intensified growth of the reserve army of labour. In contrast to the
European experience, they will not find an adequate number of jobs in the
urban sector.In the longer run, two additional factors will have to be considered. These are the sharply increasing pressure on land through the population growth and an industrialization process which reduces the price of mechanical energy in relation to human and animal energy. Both factors will
exacerbate the problem of rural underemployment/poverty,

As noted, the rural labour in most LDC's is typically underutilized, i.e.
there is not enough demand for labour to fill up the work day/year for the
average labourer.In the absence of a redistribution of land, the utilization
of this labour for capitalformation purposes, e.g. irrigation schemes, would,
in a capitalist society, be dependent upon previous taxation to supply the
wage goods needed. The political restrictions on increasing the taxes are
though very strong and, as indicated earlier, it may not be in the interest
of the urban/rural elites to utilize the labour available.

In sharp contrast stands the chinese type of rural social organisation in which the reward for extra work comes *after* its input in the form of extra output. Hence, there is very low labour cost among marginal work in progress.[35] In other words, the level of employment is not restricted by the availability of wage goods, which was the case in the wage employment system. This organizational difference opens up the possibility for using the previously underemployed labour for capital formation projects such as irrigation schemes. The Chinese commune has the further advantage that of being a system which is capable of mobilizing previously underutilized labour for *large* capital formation projects which would have gone the limits of even a large cooperative or village. The change in organizational form may in this way *open up* possibilities for increasing labour productivity which were not feasible before. In combination with the increased labourproductivity stemming from the increased land/labour ratio (which is the result of the social transformation leading to another form of social organisation) the organisational change would increase labourproductivity.

The possibilities that are opened up for increasing labour productivity may lead and has indeed led to a relative shortage of labour which has induced a process of *selective* mechanisation which has released labour from low productivity jobs. For example, an increase in the total labour supply available for capitalformation projects in rural China was made possible partly by mechanizing very labourintensive processes like grain milling, treshing and spinning. The contrasting reasons for mechanisation and the subsequent distribution of the benefits from the mechanisation between the Indian and Chinese cases are revealing. Whilst in India, mechanisation has taken place in spite of a severe unterutilization of the existing labour force, the Chinese have mechanized due to a relative shortage of labour. In the latter case, mechanization has meant that the displaced labour has been able to find employment in other activities inside the commune, e.g. rural industries, services and capital formation projects. This process of shifting labour to other activities inside the commune has been simplified by the control at the higher levels in the commune of the capitalaccumulation. This centralization of the control over accumulation also facilitates an even distribution of income between the different production teams which is of considerable importance when further mechnization of agriculture in undertaken, for example, as a consequence of introducing two or three crops anually.

171

LIMITATIONS OF THE PRODUCTIVE POTENTIAL OF THE RURAL SECTOR AND IMPLICATIONS
FOR INTERSECTORAL INCOMEDIFFERENTIALS

Whilst it is apparent that the question of institutions or social organisation
is critical in analysing employment generation in the rural sector, it is
dangerous to overestimate the absorption potentials of both agriculture and
non-farm rural activities. In the Chinese case, it has been estimated that
roughly 40 million man-years are allocated to capital construction projects
and that this figure is not likely to increase.[36] Furthermore, around 26
million people are employed in rural industrial enterprises. In total then,
66 million people are engaged in non-farm rural activities. This figure has
to be compared with the total increase in population since 1957 when these
non-farm activities really took off in magnitude. A rough estimate of the
magnitude of the population increase is 300 million. With a participation
rate of 42%,[37] only about half of the population increase could be absorbed
by non-farm rural activities. Therefore, it follows (since the urban sector
is more or less stagnating in terms of population) that agriculture has had
to absorb an increasing number of people. Despite the institutional changes
undertaken in the Chinese rural areas, this pressure on the agricultural
sector has led to a decreasing marginal productivity of labour. According to
Rawski,[38] the agricultural employment has increased from 231.5 million man-
years in 1957 to 310.3 million man-years in 1975 whilst production per man-
day in Yuan has decreased from 1.3 to 1.08 (to 0.99 in a less optimistic
estimate). Thus, whilst the rural sector may absorb the labour force, the
price for it is a very low labour productivity.

Even though the Chinese, through the Commune system, take care of the minimum
needs for their population, there are still great potentials for increasing
social tension in the future. As a consequence of our basic argument that the
industrial sector is an inherently high productivity sector (with which the
Chinese apparently agree) whilst the rural sector would have to be a relatively low productivity sector if the labour are to be absorbed there, we would
expect that an important *intersectoral distribution of income* problem will
occur.

In the general case, this problem arises from the basic fact that as income
increases, the demand for agricultural produce will become relatively smaller.
Hence, the industrial sector will allocate a constantly decreasing portion of

its productivity increase for buying the produce of the agricultural sector. This means that the trade between the two sectors (as the rural sector can only buy as much industrial goods as they can sell agricultural goods). Consequently, there will be relatively less industrial goods traded which can distribute the gains of that sector's superior productivity increases to the rural sector.

This will mean that as the industrialization process continues, and the agricultural sector is charged with the absorbtion of labour, the political problem of how to transfer 'values' from the high productivity industrial sector with perhaps 10-20% of the work force, to the relatively low productive agricultural sector, will take on increasingly stronger dimensions. [39]

In the shorter run, this problem is aggravated by the fact that the falling productivity of labour in the agricultural sector causes serious problems as the expanding industrial sector demands an increasingly greater surplus. In the Indian case, we saw that the urban sector responded to the surplus problem with strengthening the larger landowners which produce a larger marketable surplus. The Chinese have apparently responded to this very same problem by working out a three sector national development strategy. This includes an urban-based industrial sector which strives to use the most up-to-date technologies, a relatively high productive agricultural sector - mainly surrounding the urban centers - which serve the urban areas with agricultural produce, and a third sector of very lowproductive communes with both agriculture and rural industries.

If the Chinese are really serious about implementing a three sector economy, the intersectoral distributional problem will grow very serious very quickly as a few selected pockets of agricultural production units (state farms) will be given preferential treatment in terms of agricultural inputs produced by industry whilst the rest of the rural sector will be relatively isolated from interactions with the industrial sector. Thus, the choice of labourintensive technologies in the rural sector have not only created a distributional problem but also a surplusproblem. Furthermore, the solution of the latter problem (both the Indian and the Chinese solutions) exagerate the former one. Whilst it is apparent that the solution to the distributional problem must go along the lines of separating productivity from income, it is interesting and alarming to note that the Chinese say that, even though they are aware of the problem, they do not have the means of resolving them.[40]

CONCLUSIONS

In this essay, we have formulated a series of hypotheses which we argue should
be of considerable interest to organisations dealing with the longrun implications of the application of science to production.

1 The skew international distribution of R&D resources is not the main
   cause of the increasing capitalintensity of industrial processes.
2 To the extent that labourintensive industrial technologies are technically feasible, there are very strong international and external non-
   technical (institutional) limitations to the development and implementation of such technologies.
3 Even if these institutional limitations are overcome, the vast majority of the labourforce in LDC's will have to be permanently employed
   in other sectors of the economy or be unemployed.
4 The absorption of labour in the rural sector depends primarily on the
   institutional setup, the social organisation in this sector.
   The relation with other sectors is also important.
5 Whilst the rural sector can absorb the remaining labour force, this
   is done at the price of very low, and possibly decreasing productivity of labour. The existence of sectors with radically different
   labour productivity will likely lead to vast intersectoral income
   differentials. The political problem of solving this distributional
   issue will be one of the key issues for developing countries to
   deal with.

REFERENCES

1 *e.g. Meier (1976)*

2 *Sources : Employment - ILO except for Brazil. For Brazil Boletin do
   Banco Brasil (1979). Output - Un Statistical Yearbook and UN Year Book
   of National Accounts Statistics (1977). The years covered were for Output
   respectively Employment : Philippines 1960 - 75, 1960 - 75. India 1960 -
   76, 1961 -75, Korea 1963 - 77, 1963 - 77. Kenya 1967 - 75, 1967 - 75.
   Peru 1963 - 72, 1963 - 72. Brazil 1966 - 75, 1967 - 75.*

3   *Sources : (1) Table I, (2) Morawetz (1974) table I, (3) World Bank (1977) (2) was from 1970. (3) was from 1970 - 75 .*

4   *Jones (1978)*

5   *Engineering Academy of Science (1979)*

6   *Freeman (1978)*

7   *Clark (1979)*

8   *Ibid*

9   *This is confirmed by Hamilton (1979)*

10   *Skorov (1978)*

11   *Stewart (1978)*

12   *Indian National Paper (1979)*

13   *Wilton (1978)*

14   *Marx, vol. 1, chapter 13-15*

15   *Forsyth et. al. (1977)*

16   *Dickson and Marsh (1978)*

17   *Bessant (1979)*

18   *Dickson and Marsh (1978)*

19   *Rhee & Westphal (1977)*

20   *Ibid*

21   *Sercovich (1974), Mytelka (1976)*

22   *Ure (1835)*

23   *Personal communication with Dr. R. Rotwell at SPRU at the University of Sussex*

24   *Mirow (1977)*

25   *Edquist and Edqvist (1979)*

26   *Todaro (1971)*

27   *World Bank (1972)*

28   *Stewart (1978)*

29   *Junakar (1976), Rajaraman (1975), Dasgupta (1976)*

30   Raj (1972), Vashinta in Sen (1973), Sen (1973), Sen (1973), Rao (1975)

31   Dasgupta (1976)

32   Raj (1972)

33   Rao (1966)

34   Dasgupta (1976)

35   Sen (1973)

36   Sigurdson (1979)

37   Personal communication with Jon Sigurdson at the RPI at the University of Lund

38   Rawski (1978)

39   Charles Edquist drew my attention first to this problem

40   Personal communication with Jon Sigurdson

STATEMENTS

To think that unemployment can be discussed in terms of (the Lack of) appropriate technologies is basically a mistake for three main reasons :
1 the magnitude of the problem
2 the technical potential for labour intensive techniques has been
   exaggurated
3 institutional constraints for development and implementation of
   such technologies have been largely ignored.

The inherently capital - intensive character of technical change in the manufacturing sector will cause wide intersectoral productivity - differentials.
How to avoid these to become reflected in income differentials is the key -
issue in development.

The vast majority of labourforce in developing countries will have to be absorbed in the agricultural sector and can not be employed in the industrial
sector. The realisation of this will be mainly a matter of social organisation
and not of introduction of technology in the sense of hardware.

OPTIMIZING THE APPROPRIATENESS OF TECHNOLOGIES

Jean-Max Baumer*

Why not simply take any existing technical solution ?

For most production purposes there is already a technology in this world. Why invent new ones, why adapt old ones?

It all starts with a normative problem: the definition of what we call "development". Economically, each country has a given investment-volume, determined by savings. Saving will only occur when people produce economic surpluses which they do not consume. The impediments to savings may arise by force of circumstances (minimum standard for physical reproduction) or by force of socially imposed consumption-standards. An economically very poor country will show a smaller investment-volume in absolute terms of value as well as a smaller fraction of total production going into investment compared to a rich country. If the poor country decides politically, to use the limited investment-volume for implementation of the same technology that is implemented by rich countries, then it will axiomatically have a concentration in its factor-allocation; the country is unable to serve all of the people at the same time with investment, simply because there is not enough investment-volume. The society is splitted into a (usually small) group, enjoying the expensive technological standard of rich countries and a group which is deprived of investment and therefore development-wise postponed to next generations. The theoretical argument here is that capital accumulation takes time.

The question is whether this time-consuming process could not socially be better bridged by more equally distributing the limited investment-volume.

178 ** S.K.A.T., St. Gallen, Switserland*

Instead of giving a lot of investment to few spots of production, expecting a high productivity and therefore a per-capita production of, let us say; A, one might give a few investment to many spots, accepting a lower productivity of more people, also resulting in a per-capita production of A.

Taking into consideration that poor people live today and not when capital-accumulation has progressed, I feel, the limited investment-volume should be wider spread now; this implies cheaper and smaller production technologies. They do partly exist, but partly they must be developed. This statement does not mean that each and every production-process needs a "new" technology.

Coming back to the definition of development we have accepted at SKAT (Swiss Center for Appropriate Technology) a definition proposed by the Indian Pro-fessor A.K.N. Reddy: development is based on three principles :
a Fulfilment of material and non-material human needs, starting with
  the needs of the neediests, balancing out the disequilibria within
  and between nations.
b Orientation towards the proper forces of society through broad social participation and control.
c As large a harmony with the natural environment as possible.

Technology should serve these three principles. This requires that technology is adapted to a number of criteria; technology must positively respond to - as we suggest - the nine following elements, which represent a more specific des-aggregation of the above mentioned three principles. Based on the given concept of development we want to further qualify the relationship between technology and the following criteria:

Technology and Autonomy : Appropiate technologies can and must as far as pos-sible be invented, developed, constructed, produced, handled and maintained in the developing country itself. The principle of as broad a local social participation and control as possible applies here too; namely the establish-ment of particular innovation centers which are to lessen the dependencies and the influences in the technical field and allow the population to get the benefits of the manifold learning effects of the development of technology. Through self-help and local autonomy there emerges a technology which enables, strengthens and promotes a local, autonomous social development.

Technology and Culture : Appropriate technologies should be consistent with

179

culture and take given local ways of thinking and behaving into account. They try to take advantage of these ways of thinking and behaving as starting points for further development and try to cause soft cultural changements - which can easily be absorbed by the local population - with regard to a wanted social evolution.

Technology and Basic Needs : Appropriate technologies (adaptation and local development) have as their primary aim to meet the local basic needs. Moreover should they be mobilized to meet higher needs.

Technology and Economic Growth : Appropriate technologies try to set an economic growth-process going, which integrates as large a part of the population as possible and allow them to get the respective benefits. This broad effect should take priority over the concept of high growth-rates of production.

Technology and Manpower/Employment: Appropriate technologies should preserve existing employments and create new ones, which are physically acceptable and make sense. Of particular importance are thereby a low capital expenditure per place of work, no overstrain of the existing or to be acquired educational levels, no too far-reaching specialization and an adjustment to locally disposable manpower as far as this is possible.

Technology and Productivity: Appropriate technologies must be economically profitable in the sense of positive relation between cost and earnings. However, one must not only pay attention strictly to the project-profitability but also take into account negative externalities like e.g. destruction of traditional jobs, inflicted social cost of needed infrastructure etc. as well as positive effects like decentralization of production of preservation of not renewable resources. Quantitative as well as qualitative factors must be included into the cost-benefit-analysis. Such a comprehensive profitability-calculus must also be applied when the competitiveness of technical options is evaluated.

Technology and Decentralization : Appropriate technologies should help to avoid the concentration processes which are characteristic for the Industrial Countries. They should as far as possible operate at costs that can be financed by local savings and promote businesses which produce according to the size of the local markets. Appropriate technologies enable in such a

way a broad distribution of the property of the means of production and are
also particularly suited for small local self-help organizations.

Technology and Ecology: In order to protect the ecological cycles, appropriate technologies try as far as possible to use re-newable resources and to
economize non-renewable resources; they try to keep the pollution of the
environment on a low level and to work for an extensive re-cycling process.
Local resources are relevant for the choice of the appropriate technology.

Technology and Politics: Appropriate technologies will generally be the more
effective, the better the decisions with respect to the economic and social
policies are attuned to them. The competitiveness of appropriate technology
while it is being introduced, must be supported by a corresponding economic
policy of protective and stimulative measures under full consideration of
existing economic conditions or distortions. In doing so one should be aware
of the fact that appropriate technologies can only represent one - though
an important - means to attain the goals of development.
Consequently the appropriateness or inappropriateness of technology is "relative"; it is relative to the state of choosen variables, which are recog-
nised as relevant. If one assumes that the set of variables is more or less
the same like culture, ecology, manpower etc. for most countries, there will
be n-states of each variable and m-possibilities of combinations. Theoretically this would imply a indefinite number of needed technologies since
every case will be somehow different from the other. By talking about "optimizing" in this paper, I mean that the reality will be in between the
"I-Solution"(mostly from developed countries) and the "nm-Solution" described
above.

All in all we may say that the "relationships" constitute the degree of appropriateness. The technology itself may differ very much from case to case as
long as it serves the defined goals spelled out in "relationships" between
technology and relevant variables. Consequently there is no such a thing like
a worldwide "catalogue" of appropriate products and inappropriate products.

Why is "appropriate technology" not widespread?

As logically as all this seems and as old these thoughts might be, the concept
has not yet come off with flying colors. Why? Let me catalogue twenty-five

reasons which explain the tendency to modern technology instead of optimizing the appropriateness of technology.

Reasons concerning the decision-level of goals and strategies:
a the myth of productivity, the will to be efficient
b the will to possess a product-supply comparable to the one of rich countries
c the production-structures oriented towards monetarian markets like cities and exports
d the integration into the worldwide, ever more specialized division of labor

Reasons concerning the psychological level:
a the image of modernism, progress, prestige and fascination whereas AT seems old-fashion, technology of the poor.
b the mentally paralysing effect of the advanced technology which many people mythify.

Reasons concerning the development-level of AT:
a the mighty informationflow about modern technology
b the instant communication among producers and users of modern technology
c the disappearance of traditional technologies
d the inadequacy of traditional technologies due to a change of variables like e.g. population growth
e the hiding of older technologies to protect newer ones in the market
f the lack of "innovation-centers" in developing countries
g the lack of time to develop new technologies because basic problems call for immediate relief
h the ever-present offer of "drawer-technologies" being a "comfortable" way of solution
i the impediments through patents
j difficult R&D-cooperation with industrialized countries

Reasons concerning the implementation-level:
a implementation-problems in foreign-controlled and national industries in developing countries

b the mostly centralized political power in developing countries,
   not favouring regional solutions
c the economic policy of increasing labor cost and subsidising
   capital-investment, fostered by factorprice distorting capital-
   aid
d the competition with large-scale producers

Reasons concerning the diffusion-level:
a the irreversible structures like big cities, big industries,
   infrastructures, absorbing most of the investment
b the elites identifying themselves with rich classes
c the increasing foreign direct investments, usually the vehicle
   of modern technology
d the ever increasing demonstration-effect by communication ex-
   plosion
e the pitfalls of appropriate technologies, made very known by
   the same communication systems

One might turn rather pessimistic looking at this list of impediments; to
overcome them many social group, starting with Government, Universities and
professional schools, Industries, Inventors, International Agencies and
Voluntary groups will have to embark into the R&D and above all the application rather than "theorizing" of what should be done. The application starts
with a serious evalution of existing technical options. How can one proceed?

How to filter out technical options?

The following "filter" is basically socio-economic, not technical; it takes
care of the economics of technology. It is still a further desaggregation of
our previous nine criteria which were based on three principles of development. Whereas the nine criteria defined a broad frame for technology-under-
standing, this "filter" is oriented towards the operationalization of the
"choosing-process".

This "filter-system" could ask five main questions with reference to three
variables which are each to be further described normatively. The earlier
defining-description of relationships between technology and nine criteria
serve here as a guide. The "filter" would have to be superimposed by a

variable to be called "cultural comprehension" (status quo) as well as one describing goals and wanted social changes (desired quo). Out of these two results what might be called the "definition of needs". The latter would be confronted with available technical options, analyzing from step to step the effects on the relevant variables resp. their normative meaning. A feedback to the level of "status quo" is necessary after each step. Needless to repeat that no technical solution will maximize all fifteen subgoals; it is an optimizing process. It might however bring along the imagination, that no existing technique is "good" enough. Development of a new or adaption of an existing one will then become necessary.

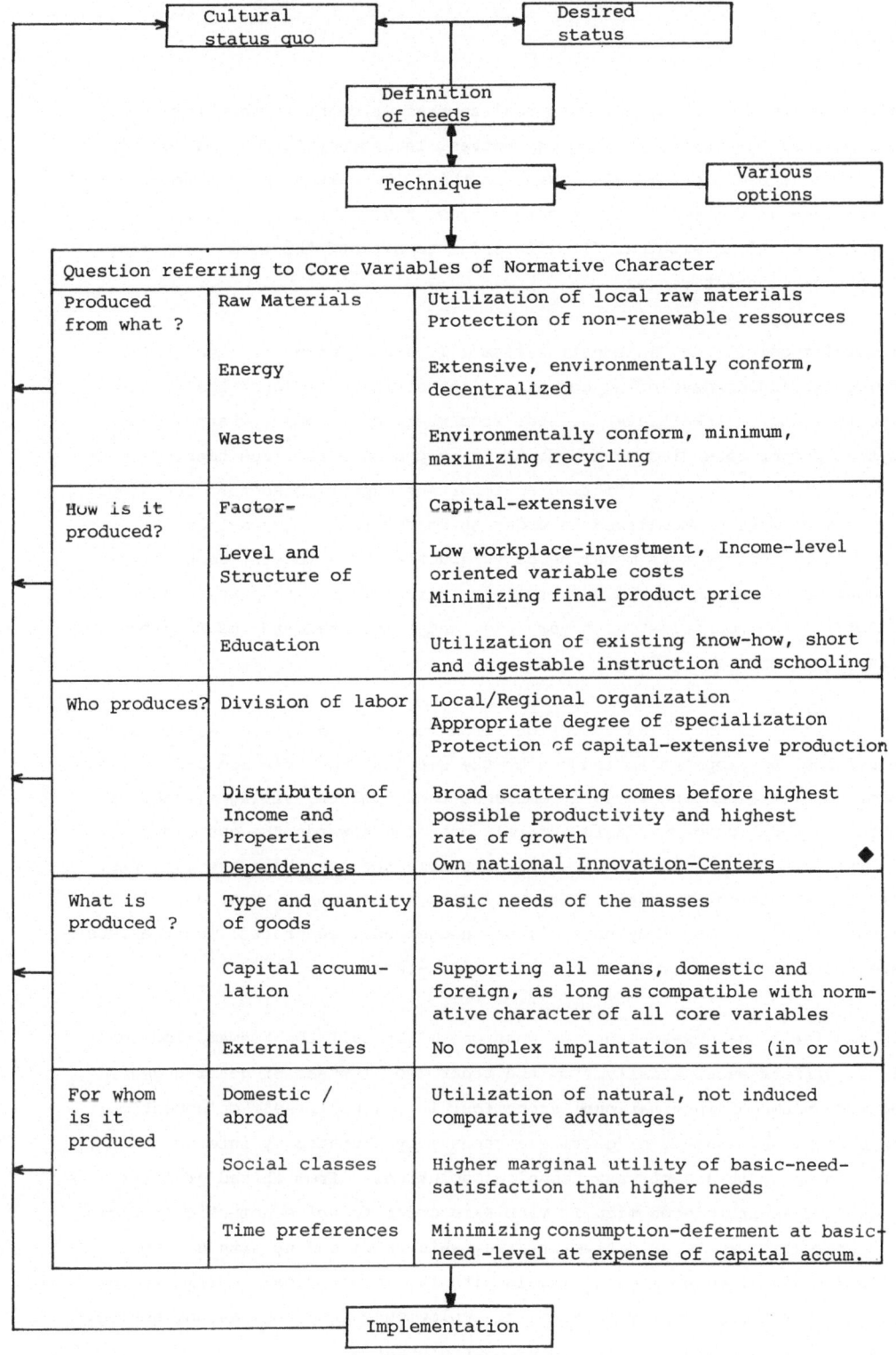

Cultural status quo		Desired status

Definition of needs

Technique ← **Various options**

Question referring to Core Variables of Normative Character		
Produced from what ?	Raw Materials	Utilization of local raw materials Protection of non-renewable ressources
	Energy	Extensive, environmentally conform, decentralized
	Wastes	Environmentally conform, minimum, maximizing recycling
How is it produced?	Factor-	Capital-extensive
	Level and Structure of	Low workplace-investment, Income-level oriented variable costs Minimizing final product price
	Education	Utilization of existing know-how, short and digestable instruction and schooling
Who produces?	Division of labor	Local/Regional organization Appropriate degree of specialization Protection of capital-extensive production
	Distribution of Income and Properties	Broad scattering comes before highest possible productivity and highest rate of growth
	Dependencies	Own national Innovation-Centers ◆
What is produced ?	Type and quantity of goods	Basic needs of the masses
	Capital accumulation	Supporting all means, domestic and foreign, as long as compatible with normative character of all core variables
	Externalities	No complex implantation sites (in or out)
For whom is it produced	Domestic / abroad	Utilization of natural, not induced comparative advantages
	Social classes	Higher marginal utility of basic-need-satisfaction than higher needs
	Time preferences	Minimizing consumption-deferment at basic-need -level at expense of capital accum.

Implementation

By talking about technology one must realize that we ought to talk first about
something else: the "thing" technology serves. Technology is never an end in
itself, it is a mean to something. What is this? It is what we call development. As crude as the models might turn out, we have to sketch briefly the
main categories of development first in order to situate the term technology
within the development context.

One general model is the "Internationalism". It starts with the idea that
worldwide, including developing countries, the division of labor has to increase, that world markets are the spot from where development starts. It
means that poorer countries should do anything possible to reach these markets
which can only be done by initiating an important exporting sector. The domestic resources must be mobilised in order to foster this outward-looking
sector. The technology this model needs is doubtlessly a modern, sophisticated technology since the world markets (highly identical with purchasing power
in industrialised countries!) set very high and fancy standards as to product
features, qualities and prices.

A reactive model is the "Disassociation" from world markets, i.e. autarkie,
self-reliance. Development is induced by the smallest possible economic units,
the family. Like expanding waves it embraces then hamlets, villages, regions
up to the national borders. It is entirely oriented towards the own end and
means. The technology needed is the only possible and feasible: the own ones:
from family-technology to national technologies at the most. Their general
characteristics are that they have all be endogenously generated. Their exists
no dependence from abroad. It is a "selfreliant-technology".

A more realistic in-between-model is what one might call the "inward-looking"
model. It differs substantially from the other two. Instead of getting geared
to the mechanisms of international economics, it orients itself on domestic
economic problems. Instead of getting a technology dictated by international
market-standards, it accepts any technical solutions - from abroad or internal - as long as it is economically viable in order to solve domestic problems.
It is a pragmatic way, a selective strategy which will end up with a complex
mix of all kinds of technological levels with the primary goal to relieve the
masses within the own country in terms of employment, food, education, health,
housing, and if possible, higher consumption-standards.

This should make clear why first of all we have to decide or find out about
the development patterns before we embark on technological decisions.

SOME TENTATIVE THOUGHTS ON DEVELOPMENT ASSISTANCE PROJECTS FOR THE RURAL
POOR, WITH THE ACCENT ON THE TRADITIONAL RURAL ECONOMY

*Camiel Baerwalt**

1   Development Assistance Projects for the poor within the traditional rural
    community - a large part but by no means all of the rural poor - nearly
    always are interventions from outside the local community. Ideally they are
    aimed at triggering off a self-sustaining social economic and technical
    development of and within that community that is congruent with the overall
    development of the country, which in most cases is heavily dependent upon
    and conform with the so called modern life pattern and its advocates.

2   With modern life pattern is meant the life pattern as it exists in the so
    called first and second world, as well as in most urban and rural growth
    centers of the third world itself, with it's inherent forms of social,
    economic and technological organisation. Modern in this context, does not
    imply a positive value judgement, nor per se a negative one. The existance
    and continous extension of this life pattern is just a political reality.

3   The means to reach the self-sustaining development of the rural poor that
    is congruent with the overall development of the country should not auto-
    matically be the change of the existing life pattern of the local community
    with it's traditional institutions into "the" modern one by the introduc-
    tion of copies or segments of the modern forms of social, economic and
    technological organisation, but rather to help the people relate to the
    modern life pattern that affects and forcefully invades their community
    anyhow. The underlying concept of this approach is that unless the local
    community is helped to relate it's development one way or another to the
    modern life pattern, community and individual life are doomed to degrade.
    Unfortunately there are already too many examples of this degration.

Eindhoven, Netherlands

188

4 Helping the local community to relate to the modern life pattern
   implies that the community should participate in the decisions on direction
   and speed of development steps to be taken.

5 The organisation of this participation and the exploration of local development needs and resources as well as of external available resources is of
   crucial and primary importance, and should be concentrated upon in the first
   stage of the development assistance intervention. Only in this way the right
   definition can be given of development potential and constraints (local
   politics !!), and only than the right decisions can be made together with
   the local community, as to nature, direction and speed of development steps
   to be taken in the next stage, the implementation phase.

6 To underline the initial openness of the intervention as to content and
   speed of the future development steps to be decided upon, dependent on the
   outcome of the participatory decision making processes of the first stage,
   ànd to accentuate the probability of an interrelated combination of different development steps to be taken in the different sectors of community
   life, we would prefer to speak of a development assistance *program*. From
   the point of view of the local community it is a development *process*. The
   total development assistance program can - and mostly will - consist of
   different *projects*,that can range from straight forward and easily planned
   relatively simple sectoral projects to, for instance, complicated multisec-
   toral ones. Often specific content and timing stay, even during the implementation phase, dependent upon an unpredictable "click" in the learning
   process involved.

7 Those learning processes are essential not only for the target groups, but
   also for the "interventionists", be it local or foreign, individual or institutional. It would be advisable to systematize or even to institutiona-
   lize those learning processes into carefully planned learning systems, com-
   prising not only the program participants involved, but also the policy
   makers and more remotely the development research institutes.

8 The approach as suggested here tries to guarantee a very essential precon-
   dition for real development of the rural poor: the possibility of integration of the new social, economic and technological "fremdkörper" into the
   existing life pattern and in traditional institutions.

9 The concept of Appropriate Technology should be placed in this context:
   thát technology is appropriate that has a positive function in the participatory development process described. It can be mainly hardware, but it is
   never restricted to it. It can be mainly in the social or economic sector
   of community life, or - to other distinction - in health, education, agriculture or industry, but then also it is never wholly restricted to it.
   It always has implications for many different aspects of community life,
   and it should be dealt with accordingly in carefully done search and learning systems. It can be small and low cost - intermediate as some people
   call it - but need not be so. It anyway is a technology that helps people
   in it's own way to relate to the so called modern way of life that invades
   their country and their own local community, so they can keep up with it
   according to reasonable standards of living. Only one of the options, may
   be open to them and choosen by them, is a rather sudden adoption and integration of modern technology as it exists elswhere, with its concomitant
   forms of social and economic organization (and disorganization)

10 It all boils down - from the point of view of the external agent - to the
   appropriate introduction of some form of modern technology, relating the
   local community to the world of modern technology, according to
   a their own definition, based on learning and integration processes;
   b the available and free local resources; and
   c the external resources to be made available to them.

STATEMENTS

- Any activity in development aid projects should start with the setting up of a participatory process in which the local people become involved in decision making about project activities.

- Such activities must all be seen in the line of a development process. Hence it would be better to speak of development-programs, including various social sectors, instead of development projects. These programs should always be long term.

APPROPRIATE TECHNOLOGY IN RURAL DEVELOPMENT

*Sorie J.B. Bangura**

INTRODUCTION

In R.U.R. (Rossums Universal Robots), Karel Capek dramatizes the annihilation
of homo sapiens by a technological creation of man himself. Robots systematically and successfully eliminate man from the face of the earth.
Hardly a century has gone by since Capek wrote his book and the world stands
on the brink of disaster. Two plights of mankind, two extremes, can be identified both critical for the existence of man. The first of these two extremes is the production of destructive weapons of warfare. Man has almost perfected these instruments and the robots produced today are capable of eliminating life from the face of the earth, in undeniable consonance with Capek's
prediction.
The second extreme is the near absence of technological advance in the service of the poor of the earth, intended to alleviate the execessive human effort required to make any piece of endeavour a worthwhile undertaking.
Thus, on one hand it is possible today to visit the moon while on the other,
over 70% of mankind does not have enough to eat. A great majority of mankind
that falls in this latter category resides in rural areas where poverty,
ignorance and disease are a reality, however unacceptable. Some areas are
even in a more deplorable situation than others. The African continent is
currently evaluated as one of such areas.
In recent history, much effort has been made to lighten the effects of ignorance and disease on human population in the continent. It is, at least a difficult task to generalise on situations and conditions on the continent and in
the developing world as a whole. Generalizations are sometimes at best half
truths. The lack of specificity is perhaps one of our problems of the day.
It is now a well-noted fact that finding and effectively applying solutions

U.N./E.C.A., Addis Ababa, Ethiopia

to the aching problems of development require a global approach. Thus, the talk of a new International Economic order. In the African continent, one of the institutions emerging as important promoters of progress is the United Nations Economic Commission for Africa,(UNECA) which carries, among others, programmes in appropriate Technology and rural development.

It is not possible, in this short presentation, to describe fully what the UNECA does or what all its programmes are. Suffice it to state that the major objective of the Commission is to effectively assist member states individually and collectively to mobilise its human and other resources for the creation of a better world for all. Within the Commission however, there is an African Training and Research Centre for Women, (ATROW) whose mandate is to promote the fuller integration of women in the development process. The work of this Centre is the immediate and major concern of this paper. However, I am invited to present the work of the Centre "against the background of a.t. and rural development as seen through your eyes." This effort I shall make with the appeal, that I must be held responsible for anything that sounds false and/or dishonest since this can only be my personal view and not my agency's policy.

THE AFRICAN TRAINING AND RESEARCH CENTRE FOR WOMEN (ATROW)

This unit in the Social Development Division of UNECA was created in 1975 and given a specific mandate already mentioned above. In its five years of existence, the Centre has grown both in staff and the number and scope of its programmes. By design and through experience, the Centre's activities and programmes are multi-disciplinary. The following are currently the major programme areas :

1   *Training and Education*

The training and mobilization of all human resources regardless of sex is accepted as a high priority. Training is carried out in many fields such as Agriculture, Home Economics, Cooperatives, Communication, Adult Education etc., Workshops, Seminars, Study tours, Short courses are organized at the national, subregional, and regional levels. The use of institutions in special areas is rapidly growing.

2   *Research*

A consistent effort has been and is being made to collect empirical data
on the situation of women in Africa. Although an academic exercise, research is undertaken with a view to revealing the status of women in Africa and to build specific and meaningful programmes on available facts.

3   *Income Generating Activities*

The idea in this area is to improve and diversify the competencies of
African women so as to increase their capacity for making higher incomes
for self and family.

4   *Village (Appropriate) Technology*

The ultimate outcome of the Centre's efforts is to develop useful implements which could be used in improving the quality of life of families and
communities.

5   *Communication & Publication*

The Centre produces many publications some of which are research-findings,
reports, manuals etc. Others are simply for public relations purposes.

6   *Population Family-life Education*

This component of the Centre's programme is mainly concerned with all aspects of family and community life. More programme areas may be listed but
these are the Centre's major ones. It is important to note that the main
clientèle of the centre's programmes are rural populations, especially
rural women. Others include the liberation movements and the urban poor.
The Centre's major objective is the achievement of a higher quality of
life for family and community through providing greater opportunities for
women.

WHY THE EMPHASIS ON RURAL WOMEN

About 51% of Africa's population is made up of women. Roughly 80% of this
female population lives in rural areas which are predominantly agrarian. In
certain countries, women may account for as much as 80% of total agricultural
production under some of the most laborious and arduous conditions. Most of
the rural population live in substandard housing and in unsanitary conditions.
Their opportunities for formal education are still comparatively poor. Their

194

access to public utilities and facilities is most inadequate. Even where a
great effort is exerted to provide the most basic and relevant services, sometimes there are far too many bottlenecks.

For a long time, women have been one of the most uncared for and misused sub-
groups in African society. The worst has always been their lot, this notwithstanding queens, female doctors, educationists or diplomats that may be identified. Their situation has not changed much and neither have things become
better for the nations to which they belong.

The problems of most of Africa seem to start from rural areas. It would seem
logical and to some extent there is evidence suggesting that the solutions
could also be found there and if not, known solutions could best be applied
there. One of the most unfortunate dilemmas of our comtemporary world is the
rate at which tried solutions are dissolved by other interacting factors leaving the situation worse than it was before the trial. The world order affects
the lives of rural people most gravely. They know least about such a world
order, have only a little or no say in it even though their physical input
may be quite substantial in the total effort.

A incredible as it may sound, we are failing in our era, to find solutions to
the problems of rural areas not because there is a lack of theory but because
of an insufficient commitment to (a) applying the best possible solutions and
(b) a near-failure in putting any confidence and trust in the intellectual
development of rural populations.

RURAL DEVELOPMENT

Rural development is all about people. In Africa, it is about people living in
age-old traditions; people threatened by and subjected to live under inter-
relationships well beyond their state of knowledge; people viewed only as
markets on the basis of a billion shirts, a billion toothbrushes and therefore billions of waiting dollars for the daring investors. This situation is
both local and international.

As bad as the situation is, hope exists because it is becoming increasingly
evident that the poor and the rich, the fortunate and the unfortunate must
co-exist if the world should be a better place for anyone at all.

Far too many theories have been tried and far too many projects have failed
in rural areas. More often than not, the rural people who have nothing to do
with the planning of such projects are blamed for the failure. I do not intend
to enter into this old debats and I therefore hasten to leave it at that.

Rural Development must lay a maximum premium on people. One major factor for most of the failures, or for the poor progress in developing rural areas is the lack of faith that the rural person can learn. Where substantial progress has been made, it is an acceptance of the premise that an investment in improving the knowledge levels of people is more useful than the application of short lived solutions to known problems.

APPROPRIATE (VILLAGE) TECHNOLOGY IN RURAL DEVELOPMENT

Whatever terminology one may choose, it has become an accepted fact that lives of people can be better if some type of technology is made available to them. The crucial questions are what type of technology and for what? ATRCW holds the views, that rural women in Africa need the type of technology which :

a) simplifies their work and relieves them of the back-breaking operation which they must carry out everyday of their lives;

b) must save their time, time which if realized, could be spent on self-improvement and other undertakings benefical to family and community living;

c) must make the efforts and labours of women more profitable through a quantitative and qualitative increase in the final products;

d) are simple enough to acquire or construct to ensure continuous application and use;

e) are developed and produced not necessarily for the pride of invention but based on and in agreement with traditional factors.

The Centre has for nearly four years, embarked on a programme of research and surveys in traditional technologies. It has cooperated with several institutions and agencies in and outside the region in pilot projects in several African countries and the number of such projects is on the increase. Projects and studies are on in Ghana, Sierra Leone, The Gambia, Senegal, Niger and several others.

Over the last two years the feeling has grown that the use of machines, implements and tools must not be treated as an isolated effort. There is therefore a search for the best possible ways of introducing village technology within given and known contents to avoid pitfalls which have caused failure in similar endeavours in the past. An important fact in the Centre's effort is the element of training that is all-pervasive. In none of the Centre's projects or programmes the element of training is missing. It is perhaps this solid preoccupation that women are a vital human resource and that they need to "catch up" with others that must be learnt from its efforts.

Personally, it is my conviction that there is no single solution to the problems of rural areas. If appropriate technology should make a vital contribution to the lives of rural people, it can do so only in programmes that are integrated and in which due consideration is given to the systemic nature of problems and therefore, obviously, of solutions. I also feel personally, that the commercial factor which usually accompanies good intentions, could defraud such good intentions, maning the spirits of clientèle and extorting any profitable attributes of such innovations.

Whatever happens in the future appropriate technology can be integrated in almost every activity in rural Africa. Agriculture, health, water supply, housing, clothing and cars of the environment are all possible areas for developing appropriate technologies.

The field is thus unlimited, the prospects immeasurable.

CONCLUSION

The analogy of Capek's R.U.R. is simply intended to compare the advances of technology in the manufacture of destructive weapons and the contemperary meagreness of developing technologies for the improvement of the quality of life for the greater fraction of the human population. That Capek's prediction could become a fulfilled prophecy is not anything to doubt. Nevertheless, it is essential for man to consider exploiting his present resources to improve the status and situation of man.

In Africa, the improvement of the status of man is greatly dependent upon woman, who are doubtlessly the axis of life patterns at the family and community levels. It is imperative to consider women as a vital human factor without whose edification, little can be achieved in rural Africa. In all rural development efforts, the central focus should be on people regardless of the nature of programmes and projects. The people -centred orientation should be the overriding element which alone can assure acceptance, continuity and progress.

As appropriate technology becomes more acceptable and picks up momentum, it should be realised right from the start that :

a) it must not, in practice, be applied in isolation but integrated into existing programmes or planned ones;

b) it will offer a golden opportunity for commercial enterprises whose interests are profit accumulation which in turn will doubtlessly render disfunctional efforts;

c) training should form an integral part of promotion of rural technologies. This type of orientation will remove the emphasis from the development of contraptions alone and give a human dimension to the discipline;

d) although, there is talk of transfer of technology, serious thought should be given to the development of technologies based on local materials and conditions

Rural development has received many years of attention. Many theories, programmes and projects have been tried with a lot of success and a lot of failure. In Africa, examples abound in quantity. The lesson has been learned that a premium must be placed in the transfer of knowledge which allows the acquirer to adjust to changing situations. Where there has been a partial development of the human resource based on sex, predominantly eliminating women, special efforts need to be made to provide opportunities for the disadvantaged.

Appropriate technology has a relevant place in rural development in Africa. The ECA is thus laying a proportionate emphasis on the matter and efforts are being made at the micro and macro - levels to promote development in Africa. Whils technology receives the attention it deserves, it must be borne in mind and constantly, that as much faith should be given in development of the intellect of rural clientèle as is given to technology itself.

STATEMENTS

- The main reason for analysing traditional technologies is that it is an attempt to ensure that when something is developed it is suitable for that
  situation.

- If we talk about rural development we have to realise that we are talking
  about a population consisting of women for approximately 51%.

- The involvement of the population is essential for appropriate technology.
  There can not be a question of relevance for a.t.-projects; there is only
  a question of continuity and possibility and probability that people will
  be able to solve problems that come up with the introduction of a new type
  of technology.

- Development may be synonymous to quality of live and this means a lot. It
  could mean infrastructure, provision of houses, availability of services,
  formal education of children, medical health, etc.. In general it means
  the best you can do for a family to live a decent life.

APPROPRIATE TECHNOLOGY   "FASHION OR NEED"

*Willem Riedijk**

In the Netherlands there is since 1978 a law to stimulate industrial investments for the benefit of economy, (1) especially for employment generation,
cleaner environment and energy conservation. In 1978 250 million guilders and
in 1979 2.500 million guilders are spent in accordance with this law. In four
years a total sum of 13.000 million guilders will be spent. (2) What happens
to this money? There seems to be no sign that more has been reached than the
stimulation of labour extensive investments most probably by bigger enterprises, which can afford to employ experts for obtaining subsidies from government, an often complicated process.
Could this be a symptom of the situation?
Another dutch example. If a farmer wants to explore the possibilities for
ecologically sound methods it will be difficult, if not impossible for him
to get money from the bankers who still believe that "big is beautiful".
If a farmer decides to start a bioindustry or any other monoculture, than he
might get the money he needs for it, on conditions decided by the banker.
A conclusion that may be drawn from this is that there does not seem to be
much space for appropriate technology in industrial countries. Relatively
few a.t. institutes are specifically working in and for industrialised
countries, like the Institute for local self-reliance in Washington. (5).
This does not necessarily mean that there is no need for a.t. in industrialised countries, on the contrary, but I think it is not wrong to say that a.t.
at the moment is mostly developed for developing countries.
An interesting general question arises from the preceeding text:

1.   "Is there any need for appropriate technology as a separate concept
     of technology?"
     This question may be split up in three other questions :

C.A.T., University of Technology, Delft, Netherlands

2.	"Do we know what people really need expressed in terms of appropriate technology?"

3.	"How can we make a.t. organisations appropriate to fulfill the real needs of people?"

4.	"Shouldn't every a.t.-project start with and be implemented by the people who need it?"

These questions may be classified as follows:

question 1.	refers to the relevance of a.t. in general

question 2.	may refer to the concept

question 3.	is about the organizational setting

question 4.	is about the implementation of projects.

1. Is there a need for appropriate technology as a separate concept of technology?

There are various possible answers to this question.
The O.E.C.D.-development center has already counted 600 units of a.t. in 1978 (3) varying considerably in size and quality. Van Brakel quotes 61 institutes of which 27 in developing countries (4).
A fact is, that there are more than a few hundred a.t. groups and institutes all over the world and that this number is growing steadily.
On the chinese countryside the largest experiments as yet with appropriate technologies in development took place in the last 20 years (6).
Apparently there is such a thing as appropriate technology. Implicitly this might mean that there is a need for it. But what does a.t. mean?
Maybe a somewhat more precise answer is the following: " in general it became clear at the Berlin Conference that the third world nations are making greater efforts to develop science and technology that are appropriate to their cultural, social and economic situation and to their levels of development. But even if the cooperation between the scientists and the rural population is successfull in overcoming the psychological barriers to the introduction of new technologies, often there still remains a certain resistance with the economic and political elites most of whom were educated in the industrial nations" (7).
The board of management of Delft University in its proposal to the university

council to start a center for appropriate technology (8) said the following about the relevance of a.t. (translation): "scientifically spoken a.t. must be placed in the field of interaction of technology and society. In this field the needs of society are formulated, technology procures fulfillment of those needs by all possible means. The reason for the growth of a.t. as a special subject in this field of interaction is the recognition that the proceeding technological development has a great number of effects on the functioning of society. A.t. gets its relevance in the organizational process to apply technological instruments for the solution of the problems of society. This means that for example if technology causes problems for society, the engineer must be offered the occasion to work on an active adaptation of technology to culture and nature." Here a.t. is seen in the context of the problems of technology and society. How is the relevance of a.t. to be judged? By the presence of a.t. organisations only, by its projects or products, or by its ideas and concepts?

2. *Do we know, what people really need, expressed in terms of appropriate technology?*

In an evaluation report (9) Nieuwenhuys describes an organisation active in the department of Cochabamba Bolivia. The group has an office in the city of Cochabamba and a fieldstation where a multidisciplinary team of physicians, engineers, agronomists etc. actually works in the field with farmers helping them according to their wishes and giving practical courses in the fieldstation they built, together with the people. The group produces easily readable documentation that gives a perspective on how to improve daily life. Documentation that helps the farmers to get a little better income. Maybe the group does not know precisely what the farmers need, but by working with them they get their confidence and consequently knowledge about their real needs.

Now, is it possible to formulate this in a general way? The I.L.O. has in 1977 published a book on "meeting basic needs". It defines food, better health care, good housing etc. as the basic needs of people.

Two interesting questions arise: Is it possible to define basic needs for people everywhere in the world and is "health care, food, housing etc." a basic need?

Maybe a basic need, I prefer the words fundamental need, is the need to live or the need to feel safe or the need to have social contacts. Then food and health care are only means to fulfill the needs. I think this separation is significant, because it enables us to see clearly that there may be different means, varying regionally to fulfill the same fundamental need. And one step further, if there are various means to fulfill fundamental needs, then the tools to produce the means to fulfill needs can be varied according to the local culture and nature.

The translation of fundamental needs into material means and the tools to produce these means could be a process to be decided upon by the people according to their culture and environment. (11)

Now, even in the western world there is lack of evidence that fundamental needs of people are fulfilled. We live, but the quality of food is growing worse every day. Even inside our houses we don't feel safe anymore. Health care is getting very expensive, but also very ineffective.

Cultures, religions, political structures, climates are so different all over the world, but has not everbody the same fundamental needs?

Assuming that this is the case, could we then develop a methodology that enables us to assess fundamental needs to translate these together with the people having those needs in terms of means and tools?

By working with the people in their situation, maybe fundamental needs can be assessed. By carefully translating those needs into means and finally into the appropriate technical and organizational tools to produce those means, a good description of a.t. may be obtained, but how could it be produced?

3. *How can we make a.t. organisations appropriate to fulfill the real needs of people?*

In the preceding paragraph the example of a development cooperation project in Bolivia showed something of the necessity to work with the people in order to know their needs. Maybe the only thing a scientist is allowed to do in a.t. is to make people conscious of their needs and the possibilities to produce means with the appropriate tools.

Supposing that this has been done and achieved. Then organizational tools and means maybe requested by the people to produce what they need.

We are still trying to solve technological problems of our society by the same methodology as, say, fifty years ago: The feasibility of a technological solution is still decided by its profitability within the existing framework of laws and regulations.

Another antiquity seems to be the scientific and technological research orga-

nisation that has been developed parallel with the industrial system.
It leaves little space for the development of alternatives. So the tools
and means offered within the existing scientific hierarchy might not be
adequate to develop technologies appropriate to fulfill fundamental needs.
A separate organizational channel might be indispensable to effectively
transfer appropriate technologies. Let me call this a transfer chain for
appropriate technology. The joint review mission of the D.T.C./TOOL project
in Indonesia (12) gives a key significance to the field station in this
transfer chain.
It should have the following functions:

"1. A basis for field activities, such as field testing in technical,
     social and economical aspects.

 2. A shop for appropriate technology in which villagers, artisans, develop-
     ment workers of government and private organisations can find what is
     currently available.

 3. A place for training for a.t.. The training is meant for villagers,
     artisans and development workers."

These fieldstations can be seen as the beginning of the chain. The fieldstations or technology shops would need backing by workshops and centers
where models can be built and tested, where courses can be designed and
prepared etc. The centers would need to be connected to - or have relations
with - research and development institutes generally of universities in the
country and abroad, where systems can be designed and more fundamental
backing can be given to the actual work in the field.
It may well be that this transfer chain: technology shop-workshop-center-
university laboratories is indispensable for a productive a.t.-organization.
The subjects, the contents to be transferred by the chain can be based upon
the needs-means-tools assessment of the villagers.
In the case of a.t. organisations, the capability of the transfer chain to
produce the technical and organizational models that can be used by the
villagers to fulfill their fundamental needs may be the yardstick for
appropiateness.

4. *Shouldn't every a.t.-project start with and be implemented by the people
    who need it?*

On first sight a trivial question. Looking at the organizational aspect of

implementation of a project, it may be clear that often the poorer people, villagers, people living in urban slums, do not have access to any financial or other means to organise themselves around concrete technical solutions for their needs. So the end of an a.t.-organizational chain can be seen as the beginning of the implementation. For the implementation of an a.t. project a detailed study of the total system to which the project belongs, will be required. During a multidisciplinary binational feasibility study on ayurveda, the traditional medical system of Sri Lanka, it was observed that:
"Ayurvedic medicine has very deep roots in the society of Sri Lanka. It was born from the indigenous "science of life" and it developed in natural lines with the population. People have an innate feeling that ayurveda has an answer for their ailments and traditionally they seek relief with their ayurvedic practitioner, who often is an influential man in the community."

One of the problems the team met (pag. 31):

"A very serious problem is presented by the ayurvedic drug supply. No open market with sufficient genuine easily available drugs is in operation. Various modes of purchase are being practised to provide the wanted medicines. Procurement of drugs from factories of the ayurvedic drug corporation is possible. Many practitioners though prepare drugs themselves or buy them from collegues. In rural areas patients receive instructions how to prepare their drug at home. In all these cases there is no guarantee for genuity and uniformity, since nowhere qualitycontrol or standardised production can be formed."

The ayurveda system belongs to the people of Sri Lanka. Real implementation may only be possible if the a.t. project, in this case improvement of the quality of drugproduction, is part of an already existing indigenous system. In any implementation problem the technique may only form a small part of the problem. The appropriate organisation and the developments necessary to create the circumstances for the existence of an appropriate organisation, may be of much greater weight than the technical problem.

REFERENCES

1. W.I.R. "Wet Investerings Rekening", Staatsdrukkerij, Den Haag, 21 juli
    1978.
    Law of june 29th, 1978 containing rules for the stimulation and regula-
    tion of the investments.

2. Vrij Nederland, 25 augustus 1979, nr. 34 page 1

3. Review of development cooporation"- O.E.C.D., 1978, p. 53-67

4. "Chemical technology for appropriate development", J. van Brakel, Delft
    University Press, 1978, page 132

5. "Appropriate technology in the U.S.A. and Canada", W. Riedijk, Delft
    University of Technology, march 1979, page 12. 13

6. "Appropriate technology for developing countries", W. Riedijk (editor),
    Delft University Press, july 1979, page 229

7. "Science to serve human needs", Hans Peter Fuhrman, Development and
    Corporation 4/79, page 21

8. "Proposal for the foundation of a center for appropriate technology",
    Board of Management, Delft University of technology, March 21, 1978,
    page 2

9. "Evaluation report on three cofinancing projects in Bolivia", M. Nieuwen-
    huys, NOVIB, The Hague, december 1978

10. "Meeting basic needs", I.L.O., Geneve, 1977

11. Course handbook "appropriate technology for developing countries" part I:
    Introduction to appropriate technology, W. Riedijk, Delft University Press,
    1979

12.   "Interim report of the joint review mission D.T.C./Tool, Herudi, Hommes, Sapii, Riedijk, Bandung, july 1979, page 27

13.   "Appropriate technology for Ayurveda", report on a feasibility study by Edirisooriya, Gunawardhana, Labadie, Nijhuis, Riedijk, de Silva and Wichremasinghe, Colombo/Utrecht, march 1978, page 32

STATEMENTS

- Every activity in appropriate technology is part of a system; be it
  ecological, cultural or even technical. Hence every activity in appropriate technology should start with an integral study about this system.

- For the development of appropriate technology there is a need for a
  technology chain to transfer innovations to the people. This could be
  done by the use of technology shops; workshops which can show equipment
  to the people and where the people can get help to organise the production of this equipment.

List of Participants

Camiel BAERWALT

Nachtegaallaan 9
5613 C.M. Eindhoven
Netherlands

Soric J. Bai BANGURA

United Nations Economic Commission
for Africa
African Training and Research Center
for Women
P.O. Box 3001
Addis Ababa
Ethiopia

Jean Max BAUMER

Schweizerische Kontaktstelle für
Angepasste Technik
Varnbüelstrasse 14
CH-9000 St. Gallen
Switserland

Chris BERTHOLET

Tilburg University
Bosweg 154[a]
Oisterwijk N.B.
Netherlands

Ben van BRONCKHORST

O.T.A. - 33
Jalan Sederhana 7
Bandung
Indonesia

Roberto CACERES

Centro de Estudios Mesoamericano sobre
Technologia Apropiada
8a. Calle 6-06, Zona 1
Apartado Postal 1160
Guatemala City
Guatemala

Marilyn CARR

Intermediate Technology Development
Group Ltd
9 King Street
London WC 2E 8 HN
United Kingdom

Jan Willem v/d EB

TOOL Foundation
Mauritskade 61^a
Amsterdam
Netherlands

Tom FOX

Director Urban Ecology Project
New York
U.S.A.

Jorge Zapp GLAUSER

Integrated Development Center
"Las Gaviotas"
Apartado Aero 18261
Bogota

Calyanatissa A. GUNAWARDHANA

Appropriate Technology Group -
Sri Lanka
10 Pendennis Avenue
(Abdul Caffoor Mawatha)
Colombo 3
Sri Lanka

Mansur M. HODA

Appropriate Technology Development
Association
Post Box 3.11
Gandhi Bhawan
Lucknow - 226001 U.P., India

Michel van HULTEN

Euro Action Sahel Accord
Representation au Mali
B.P. 1969
Bamako
Mali

Staffan JACOBSSON

Research Policy Institute
University of Lund
Magistratsvägen 55 N 111
S-222 44 Lund
Sweden

Tom A. LAWAND

Brace Research Institute/Faculty of
Engineering
Mc Donald College of Mc Gill Univ.
1. Stewart Park
Ste. Anne de Bellevue
Quebec
Canada, H 9X 1 Co

Robert P. MORGAN

Washington University St. Louis
Department for Technology and
Human Affairs
Campus Box 1106
St. Louis, Missouri 63130
U.S.A.

Paul OSBORN

TOOL Foundation
Mauritskade 61^a
Amsterdam
Netherlands

K. Krishna PRASAD

Eindhoven University of Technology
Postbus 513
Eindhoven
Netherlands

Willem RIEDIJK

Center for Appropriate Technology
Delft University of Technology
Mijnbouwplein 11
2628 R.T. Delft
Netherlands

Anton SOEDJARWO

Dian Dessa
Kerto 8
Yogyakarta
Indonesia